INTRODUCTION TO CLINICAL ENGINEERING

INTRODUCTION TO CLINICAL ENGINEERING

SAMANTHA JACQUES, PHD, FACHE
Vice President of Clinical Engineering, McLaren Healthcare, Grand Blanc, MI, United States

BARBARA CHRISTE, PHD
Dean, School of Engineering Technology, Farmingdale State College, Farmingdale, NY, United States

Academic Press is an imprint of Elsevier
125 London Wall, London EC2Y 5AS, United Kingdom
525 B Street, Suite 1650, San Diego, CA 92101, United States
50 Hampshire Street, 5th Floor, Cambridge, MA 02139, United States
The Boulevard, Langford Lane, Kidlington, Oxford OX5 1GB, United Kingdom

British Library Cataloguing-in-Publication Data
A catalogue record for this book is available from the British Library

Library of Congress Cataloging-in-Publication Data
A catalog record for this book is available from the Library of Congress

ISBN: 978-0-12-818103-4

For Information on all Academic Press publications
visit our website at https://www.elsevier.com/books-and-journals

Publisher: Mara Conner
Editorial Project Manager: Isabella C. Silva
Production Project Manager: Nirmala Arumugam
Cover Designer: Christian J. Bilbow

Typeset by MPS Limited, Chennai, India

Dedication

For my loving husband Brian and son Christian, thank you for allowing me to always follow my dreams.

—Samantha Jacques

How grateful I am for the endless support from family and friends along this formidable journey. Sam, we made a great team!

—Barb Christe

Contents

Foreword

When I was asked to write an introduction to this book, I thought "Wow, I've been in this business since 1980 and have seen it from many sides, so I can just discuss what I have experienced." Of course, thinking deeper revealed that would be a book of its own. While many concepts are the same today, clinical engineering is very different in today's world.

The first thing you will discover is there remains a lack of consistent terminology, which still creates confusion. This extends from the name of the profession to the hospital departments to the job titles, where many technician positions are labeled engineers. Try to think past the naming conventions and focus on the type of work being done and the value that a clinical engineer brings to health care. The lack of consistency may contribute to the great differences between what a clinical engineer may do at one site versus another. There is a great diversity of focus that different clinical engineers are engaged to provide. The commonality is the engineering systems thinking process.

What you will find as you read through this book, is that one of the truly great things about being a clinical engineer is that it is impossible to become bored. A clinical engineer serves to a large extent as a bridge between technology and medicine. In this capacity, there are myriad opportunities in the ever-changing world of health care. Whether the problem to solve is urgent or strategic, or covers direct patient care, equipment integration, human factors, cybersecurity, or financial improvement, there is always a long list to improve the performance of the health system or related organization. For me, this is nirvana. After 39 years, I am as excited to see what tomorrow brings as I was when I started my first job as clinical engineer at Hurley Medical Center in Flint, Michigan, United States.

In those 39 years, I have had the privilege to work directly for large and small hospitals, an academic medical center, a medical equipment insurance company, a consulting company focused on clinical engineering, and independent service organizations. I have been through hospital mergers and acquisitions, and mergers and acquisitions in the service space as well. In addition to running small and large programs, I have served on an artificial heart transplant team, taught hemodynamic monitoring principles, served as an expert witness, made connections around the world, and served on what seems like a million project teams. What you can expect in this profession today is still only limited by your imagination.

Lawrence (Larry) W. Hertzler
CCE, AAMI Fellow

CHAPTER 1

The profession

Introduction

Healthcare delivery has evolved to depend on technology, both simple and highly complex. A diverse group of professionals supports clinicians in the utilization of devices, software, and systems to deliver patient care. Broadly, these efforts are labeled healthcare technology management (HTM), a discipline that interweaves patient safety, medical technology, and financial stewardship. These relationships and activities are illustrated in Fig. 1.1.

Clinical engineers, as part of HTM, are one of the many contributors to the delivery of safe and effective healthcare. Possessing a unique skill set, clinical engineers collaborate with clinicians of all types, technicians, facility managers, administrators, information technology (IT) support staff, risk managers, and administrators, unified in a mission to enhance healthcare.

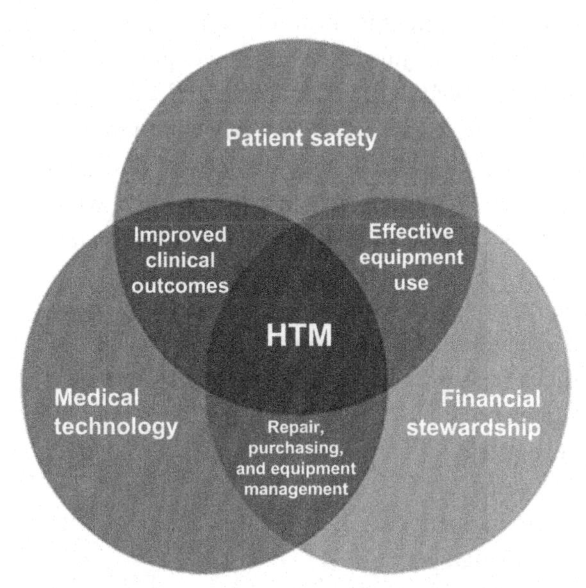

Figure 1.1 The healthcare technology management profession. *AAMI, used with permission.*

Introduction to Clinical Engineering
DOI: https://doi.org/10.1016/B978-0-12-818103-4.00001-6

What is clinical engineering?

The American College of Clinical Engineering (ACCE) defines a clinical engineer as follows:

> *A **Clinical Engineer** is a professional who supports and advances patient care by applying engineering and managerial skills to health care technology.*
>
> ***About ACCE: Clinical Engineer (n.d.)***

The *Clinical Engineering Handbook* describes the practice of clinical engineering as the application of engineering principles, such as analysis and systems principles, to improve healthcare (Dyro, 2004), acknowledging the highly complex systems of the healthcare environment featuring patients, technology, facilities, and users. Clinical engineers possess diverse knowledge and, thus, can bridge multiple domains of the environment of healthcare with technology, facilitating communication and understanding. The evolving nature of healthcare and the tools associated with patient care also drives clinical engineering as a profession. As the definition of *physician* has adapted and changed over the history of healthcare, in a parallel way so too has the definition of *clinical engineer.*

The term *clinical engineer* is not universally utilized and may be used interchangeably in healthcare settings with the term *biomedical engineer.* The confusion is exacerbated by academic institutions in the United States that do not grant undergraduate clinical engineering degrees, which can be obtained in other countries. Academic degrees are generally in biomedical engineering (BME). Many hospitals and other healthcare institutions title the position as *biomedical engineer,* although the responsibilities more closely align with clinical engineering. In addition, many departments are named *clinical engineering* but employ those with BME position titles. The converse is also true; hospital departments named *BME* also employ workers with a clinical engineering job title.

With this in mind, an exploration of the broad discipline of BME may be instructive. The Biomedical Engineering Society (BMES) offers a list of common focus areas of biomedical engineers:

1. neural engineering;
2. bionanotechnology;
3. systems physiology;
4. rehabilitation engineering;
5. orthopedic bioengineering;
6. medical imaging;
7. clinical engineering;

8. cellular, tissue, and genetic engineering;
9. bionics;
10. biomechanics;
11. bioinstrumentation;
12. biomechatronics; and
13. biomedical electronics (FAQs about BME, n.d.).

The BME profession encompasses diverse specialties including artificial limbs and cellular-level research. The many applications of engineering principles associated with BME are far larger than the more narrowly defined area of clinical engineering. Thus, clinical engineering may be considered a focal area within the wide spectrum of the BME discipline.

The World Health Organization (WHO) also utilizes the term *biomedical engineer* as a broad label for a diverse profession. In their 2017 publication, the group noted the variations on the term:

> *"Biomedical engineering" includes equivalent or similar disciplines, whose names might be different, such as medical engineering, electromedicine, bioengineering, medical and biological engineering and clinical engineering.*
>
> ***World Health Organization (2017)***

Complicating professional labels is a common job title within the clinical environment: *biomedical engineer*, often a position name provided to technicians and technologists who support devices and medical equipment. These technicians generally focus on the applications of technologies, while utilizing hands-on experience and providing services such as repair, performance assurance testing, and preventive maintenance. Academic preparation varies, but many technicians in these roles have earned an associate's degree in electronics or other closely related discipline. The relationship among the technician, engineer, and leadership, as characterized by the professional society, the Association for the Advancement of Medical Instrumentation (AAMI), is shown in Fig. 1.2.

Historical perspective

In the late 1960s, the profession of clinical engineering emerged as a BME specialty focused on broad issues in medical technology implementation beyond device maintenance and repair. Hospital safety awareness dramatically surged with the publication of an article written by Ralph Nader in *Ladies Home Journal* in March 1971. Nader claimed that there were a large number of hospital electrocutions each year. Also during this period, the community hospital system was expanding, and technology

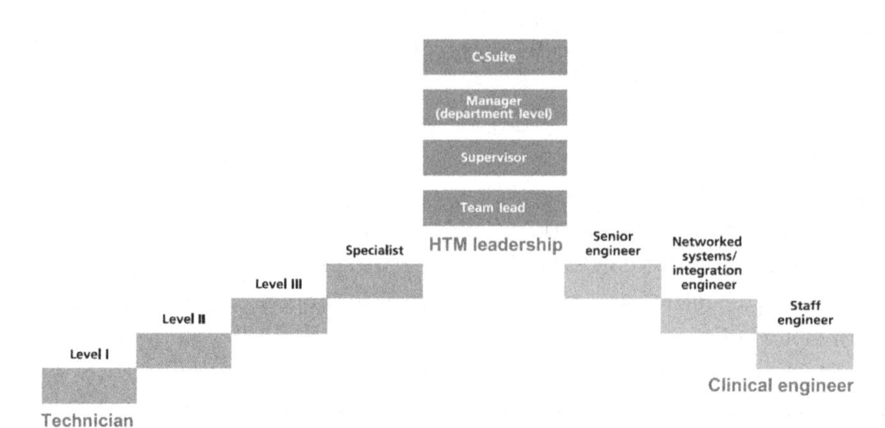

Figure 1.2 HTM career ladder.

was increasingly utilized to monitor and image patients electronically. As a result of these converging events, several organizations such as the National Fire Protection Association (NFPA) and the American Hospital Association (AHA) sought to establish guidelines and regulations to promote patient safety. In 1973, AAMI developed a certification program for clinical engineers to supplement the biomedical equipment technician certification that has been available since 1970.

Hospital equipment inventories dramatically increased in the 1980s and coincided with a societal desire to decrease the cost of healthcare. This drove an expansion of hospital-based support staff for technical services as well as a desire for technology planning and broader thinking beyond "break-fix" technology maintenance. Regulations associated with medical devices dramatically expanded with the creation of healthcare facilities code NFPA 99 in 1984, a tool that sought to establish criteria based on risk to patients, staff, or visitors in healthcare facilities to minimize the harm caused by fire, explosion, and electrical hazards. As a result, clinical engineers moved into the realm of regulatory compliance and safety advocacy.

The Safe Medical Devices Act, which requires hospitals to identify and report serious problems with medical devices, was passed in 1990. The hiring of engineers to support medical technology expanded at that time. Driving this shift was the Veterans Administration (VA) hospital system that divided the country into BME districts and hired biomedical engineers with academic degrees in engineering to oversee engineering activities.

Throughout the early 2000s, leaders in the clinical engineering profession including Joseph Bronzino, Joseph Dyro, Malcolm Ridgway, Yadin David, and many others sought to differentiate clinical engineering as an engineering discipline with an engineering scope of practice, promoting attainment of baccalaureate and master's level engineering degrees. Efforts included expansion of academic programs, improved professional publications in the discipline, and public advocacy. This drive sought to improve professionalization and status within society in general and the clinical environment in particular, defining educational requirements and differentiating the hospital technician from the engineer. Expansion of hospital roles in facilities management also occurred during this time.

A 2005 joint report from the National Academy of Engineering and the Institute of Medicine recommended expanding the role of engineering in healthcare, specifically, to apply the principles of systems engineering and IT for the delivery of healthcare (National Academy of Engineering and Institute of Medicine, 2005). The report sought to address the "growing realization within the healthcare community of the critical role information/communications technologies, systems engineering tools, and related organizational innovations must play in addressing the interrelated quality and productivity crises facing the healthcare system." The report recommended the application of engineering principles such as human factors, supply chain management, and systems modeling as tools to support quality healthcare and was viewed as a catalyst for improved recognition of clinical engineers. However, widespread recognition of the value of these engineering principles by healthcare organizations has emerged slowly since the report publication.

One dramatic leap forward in IT applications was associated with implementation of the American Recovery and Reinvestment Act (ARRA) adopted on February 13, 2009. The Health Information Technology for Economic and Clinical Health (HITECH) Act part of the ARRA offered financial incentives to hospitals to enact electronic medical records, driving the move from paper medical records to electronic files. A 2012 deadline drove major changes in the format of patient medical records, the interconnectedness of medical devices to the medical record, and the management of patient data. The role of clinical engineers in managing potentially massive amounts of patient data and coordinating the connection and communication of devices through the hospital's network has expanded additional specialty areas such as network clinical engineering and cybersecurity clinical engineering. Interconnectedness has driven a closer working relationship between the clinical engineering profession and IT department staff.

The profession

Clinical engineers are tightly integrated into the healthcare system through the support of existing medical equipment technology, the development of medical devices, and the creation of new knowledge that drives innovation in medical technology. The WHO offers a well-crafted diagram to visualize the interconnectedness of these professional facets.

The relationships as shown in Fig. 1.3 between the medical device industry, the government as regulator and funding source, and healthcare providers as healthcare delivery venues are interwoven with academic opportunities. Innovations in medical technology evolved from basic research are implemented by medical device manufacturers. A clinical engineer must have an eye toward future technologies, improvements in existing technologies, and techniques to deliver patient care effectively.

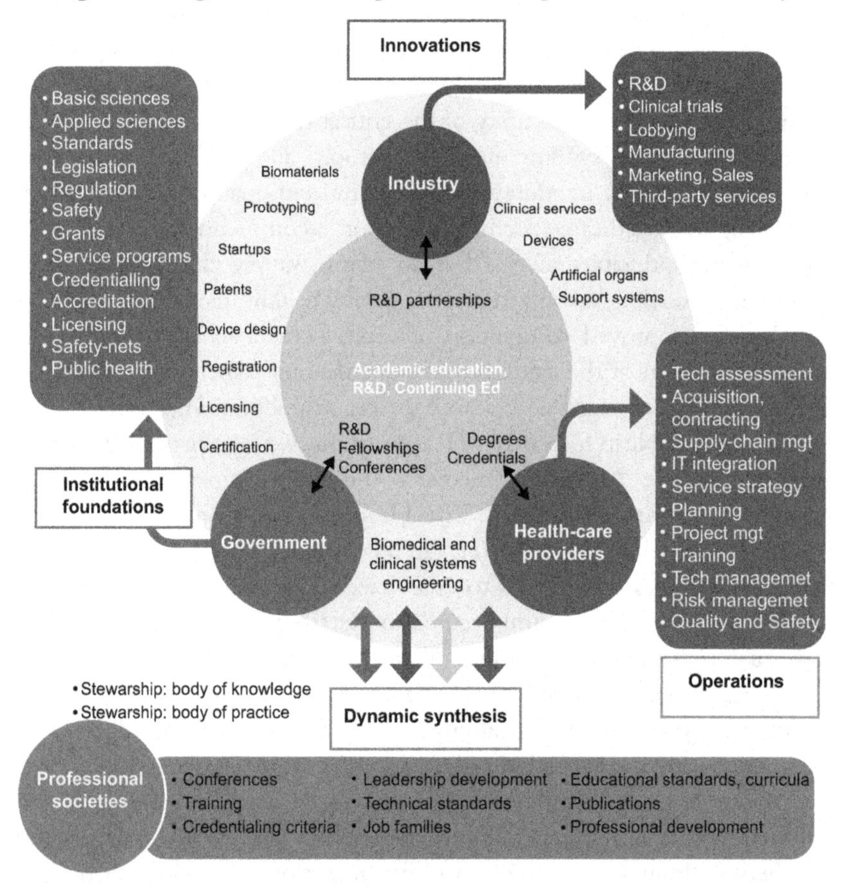

Figure 1.3 The biomedical engineering profession.

Academic pathways

In the 1970s and 1980s, BME academic programs were very limited in number or were limited to tracks available within traditional engineering disciplines of electrical or mechanical engineering until funding became available through the Whitaker Foundation. Beginning in the 1980s, the group spent millions of dollars to establish or help establish approximately 50 programs in BME across the United States. The foundation closed in 2006. In addition, academic competitiveness spurred the development of programs at non–Whitaker-funded institutions. However, all or almost all of the programs created or faculty funded focused on the theoretical research options associated with BME. Thus, while many colleges and universities offer BME degree programs, few focus on the competencies associated with the profession of clinical engineering. In addition, no institutions in the United States confer a bachelor of science in clinical engineering degree.

Federal government regulations drive some career pathways. The VA hospital system is required to hire graduates with bachelor's degrees from engineering programs that are accredited by ABET, a national accreditation organization, to fulfill engineering positions. To manage this requirement, the VA hospital system has a robust mentoring program in place to train BME graduates in clinical engineering concepts through workplace experiences.

One graduate program, established in the early history of clinical engineering and moved through several academic institutions, is currently offered at the University of Connecticut. This program features in-depth internships within hospitals throughout New England combined with coursework. Graduates are highly sought-after and readily find employment across the United States. The degree awarded is a Master of Science in BME. A second graduate degree is available from this institution, a Master of Engineering in Clinical Engineering. This appears to be the first degree in clinical engineering in the United States. Applicants are expected to have a bachelor's degree in engineering plus a minimum of 3 years of clinical engineering experience. In addition, the coursework is available for study at a distance. This opportunity may support engineers who are currently in the profession and seek a graduate degree.

Certification and credentialing

Unlike many professions within the US healthcare system, governmental agencies do not require a license or credential to be employed as a clinical engineer or to manage healthcare technology. Voluntary and optional certifications are available that offer the ability to document proficiency within a discipline and reflect a high level of competence. Employers vary as to their emphasis and/or reward regarding the attainment of certification credentials. The most common certification for clinical engineers is Certified Clinical Engineer (CCE), which is offered through the ACCE by the United States and Canadian Board of Examiners for Clinical Engineering Certification.

The purpose of clinical engineering certification, as outlined by the ACCE, is "to promote healthcare delivery improvement in the United States through the certification and continuing assessment of competency of professionals who support and advance patient care by applying engineering and management skills to healthcare technology" (American College of Clinical Engineering, n.d.). The certification process includes:

1. establishing and measuring the level of knowledge required for certification as a clinical engineer;
2. providing a standard of knowledge requisite for certification, thereby assisting the employer, public, and members of the health professions in the assessment of the clinical engineer;
3. recognizing formally those individuals who meet the eligibility requirements of the board and pass the examination; and
4. requiring continued personal and professional growth in the practice of clinical engineering to maintain certification.

Current eligibility requirements to obtain certification include one of three options:

- a professional engineer (PE) credential plus at least 3 years of clinical engineering experience;
- a bachelor's degree in engineering from an ABET-accredited institution plus 4 years of engineering practice and at least 3 years of clinical engineering experience; and
- a bachelor's degree in engineering technology from an ABET-accredited institution plus 8 years of engineering practice and at least 3 years of clinical engineering experience.

Note: the board may decide to accept academic degrees from programs in the United States that are accredited by other agencies in addition to ABET or that are not accredited. In addition, the ACCE defines clinical engineering practice as engineering practice within the clinical environment (the healthcare delivery system) or in support of clinical activities (healthcare delivery and patient care).

Note that the CCE credential existed before the ACCE oversight that began in 2002. Qualifications for certification were very different when the first group of 49 professionals was deemed qualified in 1975. As the profession was in its infancy, the individuals were named because of significant contributions to the field and workplace experience. By 1999, almost 500 individuals had earned the CCE credential through examination and review. This evolution in certification is reflected in the diverse backgrounds and training of many CCEs.

CCE certification is currently a three-step process and includes:

1. application review by the United States and Canadian Board of Examiners for Clinical Engineering Certification;
2. written examination; and
3. oral examination.

The contents of the written examination include 150 multiple-choice questions associated with the following topics:

I.	Technology Management	31.7%
II.	Service Delivery Management	19.9%
III.	Product Development, Testing, Evaluation, and Modification	4.3%
IV.	IT/Telecom	6.8%
V.	Education of Others	8.0%
VI.	Facilities Management	5.7%
VII.	Risk Management/Safety	10.2%
VIII.	General Management	13.4%

The oral examination for US candidates consists of questions related to each of three clinical engineering scenarios. Seventy-five percent of the score is based on information from the candidate's responses, and 25% is based on verbal presentation of clinical engineering ideas in an organized and professional manner. Each scenario features five to seven questions that must be answered in 20 minutes.

Once earned, the CCE credential must be renewed every 3 years. Qualifications for renewal include employment in the profession; continuing education including webinars, workshops, and academic courses; professional activities including society memberships and publications; and miscellaneous activities related to clinical engineering professional career enhancement. Renewal depends on a point system assigned to each activity. Additional qualifying activities are described on the ACCE website.

A second credentialing option is Certified Healthcare Technology Manager (CHTM), which is offered through AAMI. This certification is designed for leaders who are focused on both the management of healthcare technology operations and the management of personnel. Eligibility is broader than CCE, and options include:

- CCE certification plus at least 3 years of work experience as a supervisor or manager in the past 5 years;
- completion of the Department of Defense's biomedical equipment maintenance technician training program plus at least 3 years of work experience (military or civilian) as an HTM supervisor or manager in the past 5 years;
- an associate's degree in biomedical technology, related healthcare discipline, IT, or business plus at least 3 years of work experience as an HTM supervisor or manager in the past 5 years;
- a bachelor's degree in engineering, IT, technical writing, business, or other permitted degree plus 2 years as a manager in the past 5 years; or
- at least 7 years of work experience in the HTM field plus 3 years as a manager in the past 5 years.

The written examination consists of 100 multiple-choice questions delivered in 2 hours. Topics include financial management, risk management, operations management, education and training, and human resources. Renewal of certification is required every 3 years and requires demonstration of professional development and continuous work in the profession.

A third possible credential is Certified Biomedical Equipment Technician (CBET), which is also offered through AAMI. With a focus on technicians, this certification is designed for HTM professionals who troubleshoot, test, repair, and calibrate medical devices and networks. The CBET credential is earned through successful completion of a written examination containing 165 multiple-choice questions. Eligibility requires 2 to 4 years of work experience in the profession, depending on academic credentials. Topics on the examination include anatomy and physiology, safety and regulations, basic electronics, healthcare technology function, device troubleshooting, and IT. Renewal of certification is required to maintain the credential.

Note that ACCE, CHTM, and CBET certifications are not entry-to-practice credentials, in contrast with most medical professions, such as nursing, that require a gateway credential, typically a license. Work experience in the clinical engineering profession for a number of years is required before the credential can be earned.

Responsibilities and roles

The WHO describes the role of clinical engineers narrowly as including the following responsibilities:

- identification of institutional technology needs;
- equipment purchase planning, including assessment and procurement;
- installation of new equipment;
- management of device support during lifespan, including end-of-life disposition; and
- integration of technology with networks and the electronic medical records (World Health Organization, 2017).

However, most leaders in the profession would expand the responsibilities more broadly to include strategic planning to incorporate additional stakeholders including human capital, ongoing support, and patient safety needs. Seeking to more accurately characterize the responsibilities of the clinical engineer, the following competencies are associated with clinical engineering:

- broad understanding of the institutional technology needs associated with the delivery of patient care by clinicians;
- ability to apply systems thinking principles to improve complex processes within the institution;
- ability to apply life-cycle management principles to technology;
- ability to translate technical concepts and educate constituents;
- broad understanding of regulatory compliance requirements;
- ability to apply quality assurance practices to technology management;
- broad understanding of human capital management to support medical technology; and
- visionary and strategic thinking for technology acquisition and utilization planning.

The scope of these activities is expanding significantly as healthcare technology continues to become more complex and integrates with communication and information systems. Clinical engineers involved in medical device implementation may oversee the integration of activities among clinical systems, as well as utilize cybersecurity measures to protect medical device and patient data.

Potential employers and career pathways

Clinical engineers are employed by a wide variety of groups, yet they have similar position responsibilities. Some clinical engineers are directly employed by hospitals or affiliated hospitals. Third-party groups such as outside service organizations or independent service organizations (ISOs) may manage the HTM-related services for an institution and, thus, clinical engineers and technicians may be employed by these organizations but located within a clinical facility. Those employed by ISOs may appear the same as all other hospital employees to the clinical staff, utilizing facility email addresses and ID badges. In some larger organizations, clinical engineers share their expertise across multiple institutions. Clinical engineers may also be employed by medical device companies to help support and implement technologies in various healthcare settings.

In addition to the clinical engineering competencies described in this chapter, some professionals have a more narrowly defined scope of expertise. Clinical engineers in private practice may serve as consultants, often focused on particular facets of the profession such as cybersecurity or procurement. In addition, clinical engineers work as legal consultants offering expert testimony, as specialists for the Food and Drug Administration, as chief engineers for accrediting organizations such as the Joint Commission, or as engineers at the WHO.

Many clinical engineers emerge as leaders of departments or groups during career progression. Cultivating the qualities of excellent leadership and management can be critical to the successful completion of many of the tasks associated with the profession. Clinical engineers should develop an eye toward emerging challenges and potential accomplishments, coupled with successful strategies to promote departmental achievements. Mentorship and professional development can support the acquisition of these leadership skills.

Societies and collaboration

Professional societies and organizations offer a variety of forums to connect groups of people involved in similar disciplines to share ideas and promote a deeper understanding of the challenges and opportunities in the field.

American College of Clinical Engineering

The ACCE was founded in 1990. The mission of the group includes:

- establishing a standard of competence and promoting excellence in clinical engineering practice;
- promoting safe and effective application of science and technology in patient care;

- defining the body of knowledge on which the profession is based; and
- representing the professional interests of clinical engineers (About ACCE: Our Mission, n.d.).

The ACCE serves as an advocate for the clinical engineering profession, recognizing individual achievements through awards and acting as a conduit for information dissemination. In addition, the group offers academic scholarships, a student paper competition, as well as workshops and webinars for professional development. Webinars related to a wide variety of topics are often archived and available for purchase. Student memberships are available at a very low cost.

Association for the Advancement of Medical Instrumentation

AAMI was founded in 1967 and represents a broad group of professionals of diverse backgrounds. Members are connected by one mission—the development, management, and use of safe and effective health technology.

Activities of AAMI include:

- developing medical device standards;
- providing training programs, webinars, and conferences;
- publishing practical guidance documents, publications, and other resources;
- encouraging networking and sharing of expertise through online forums, podcasts, videos, a blog, and summits;
- partnering with key stakeholders including federal regulators, HTM professionals, clinicians, industry experts, and researchers to find solutions; and
- advancing patient safety through issue-oriented projects related to healthcare technology.

AAMI offers a major annual conference, scholarships, and several national awards that unite many professionals associated with HTM. Student memberships are free. In addition, the group has published several useful free booklets for clinical engineers including the *Leadership Development Guide* and *Career Planning Handbook*. Additional publications are available for a fee including the *Healthcare Technology Management Manual* and *Electrical Safety Manual*. Training opportunities and webinars are also available through AAMI University. These HTM-specific resources can share context-driven best practices and opportunities to gain insight from the experiences of others in the profession.

American Society for Health Care Engineering

The American Society for Health Care Engineering (ASHE) is a group related to the AHA. ASHE represents professionals who design, build, maintain, and operate hospitals and other healthcare facilities. ASHE members include healthcare facility managers, engineers, architects, designers, constructors, infection control specialists, and others. For clinical engineers who are closely involved in facilities management, this group offers education, regulatory guidance and advocacy, and networking to promote collaboration.

Local societies

Many areas of the United States are served by associations and groups that gather HTM professionals for conferences, meetings, discussions, and social outings. An extensive list of regional HTM societies is available on the AAMI website. Some prominent examples include the Indiana Biomedical Society, the North Carolina Biomedical Association, and the New England Society of Clinical Engineering (NESCE). NESCE has been established longer than most groups and serves a broad group of HTM professionals. Regional symposia can offer a cost-effective opportunity to network with other clinical engineers.

Professional development

Professionalism is commonly related to attire, body art, behavior, etiquette, and written communication skills interwoven into technical job functions. Professionalism can be distilled to 11 essential characteristics: organization, timeliness, communication skills, ability to accept criticism, empathy, demeanor, grooming, collaboration, initiative, self-improvement, and adaptability. Self-improvement embodies a willingness to seek out educational opportunities. Adaptability includes a desire to be flexible when changes occur in the workplace, including the introduction of new technology or regulations. Both are related to lifelong learning.

For clinical engineers, the expansion of technical knowledge is not just recommended but vital. Someone once generalized that everything technical that a student learns before graduation is irrelevant 5 years after graduation. Perhaps this statement is an exaggeration, but the use of technology in medicine is evolving continuously. Willingness to learn more over a professional lifetime, read more, attend conferences, and remain current is vital to a successful career as well as the safety of patients. A continuing debate revolves around the concept of lifelong learning. Are clinical engineers unethical if they do not maintain current skills? Are out-of-date technical skills unethical or simply unprofessional? Is there a difference? These important questions should drive efforts to improve technical awareness and competence.

To better understand evolving technical competencies and manage challenges, clinical engineers often participate in continuing education events or webinars that feature experts sharing ideas over the Internet. Topics vary widely from highly specific regulatory compliance to career development to looming challenges like cybersecurity. Some webinars have a fee associated with them. Often, real-time participation is not needed as an archived version is available.

Scholarly and indexed journals of the profession include the *Journal of Clinical Engineering* and *Biomedical Instrumentation and Technology*. Several readily available trade publications include *24* × 7 and *TechNation*. Both are available on the Internet at no cost. Reading journals and other publications offers the opportunity to explore technical challenges that may be shared as well as data-driven studies that offer best practices in the HTM profession.

Code of ethics

Definitions for ethics vary widely, but a general description includes the moral choices/behaviors made by people in relationships with other people. Humans have rules and standards that guide behavior choices. Ethics often refers to the norms of a society or group. Professional ethics is associated with the workplace and is shaped by three forces:

1. *Legal behavior.* Legal issues are handled by local, state, federal, and international law and are managed by lawyers and the legal system in general. Sometimes a legal action is not ethical, and sometimes an ethical decision is not legal.
2. *Code of ethics compliance.* Codes of ethics are often developed and managed by individual professional societies. For example, clinical engineers are guided by the ACCE Code of Ethics for Clinical Engineers, and electrical engineers follow the Institute of Electrical and Electronics Engineers Code of Ethics.
3. *Is it the right thing to do?* At the end of the day, when all is said and done, engineers must *never* compromise integrity or the safety of any member of society including patients, care providers, or visitors.

Professional ethics for engineers are very important because decisions can affect the health and safety of society, decisions can have financial and environmental impact, and unethical choices can have professional or legal implications. Professional ethics serve as a guide for *occupational identity*.

A *code of ethics* is a framework developed for a profession to guide ethical judgment. Typically, codes serve as a tool to articulate the ethical expectations of a profession. The parts of a code often characterize the roles and responsibilities of members of a group. A professional code of ethics is not a set of rules and does not establish new principles or guidelines. A code of ethics can establish a common form of professionalism.

Some professional codes are very short and general, or long and detailed. Professional ethical codes proscribe a certain way of acting for members of a group, to serve the interests of all members of a profession. A code of ethics can define a *value system*, identifying priorities. Most professional codes of ethics are vague, incomplete, and require interpretation. In general, codes of ethics simply describe elements of good professional practice.

The author of an article written before a code of ethics was developed for clinical engineering suggested that a hospital environment demands that "ordinary engineering ethics is not appropriate," asking "should the obligations of clinical engineers differ from all other engineers?" (Davis, 1992, p. 180). The environment of engineering in healthcare imposes several unique conditions:

- *Service to the community*. Work is related to the health and welfare of the public, as opposed to the profits of a company or the happiness of a client.
- *Patient vulnerability*. Young, old, and sick individuals are typically the most vulnerable in a community and are all served in the clinical workplace.
- *Complexity of problems*. In the manufacturing workplace, the corporate "task" is singular: to create a good product. In the clinical setting, the human body system is extraordinarily complex, and multiple diseases demand high-level problem solving. Even automobile manufacturing, which creates a highly complex product, is contained within a much simpler environment, simply because the interactions of systems are fairly well-understood and defined.
- *Financial complexities*. Who pays the bills? Who is the customer? The unique relationship among uninsured individuals, Medicare, Medicaid, insurance companies, and private-paying individuals results in a unique "business" model that complicates traditional profit-and-loss relationships.

- *Culture of medicine*. The relationship between physicians and staff is complex. The decision-making structure of the hospital is quite unique. In most cases, the physician has the ultimate decision-making authority.
- *Federal privacy regulations*. The Health Insurance Portability and Accountability Act (HIPAA) is a 1996 law that addresses which patient information may be shared and with whom. This federal law forces the compliance of individuals with patient privacy.
- *High-stakes decision making*. Do not resuscitate (DNR) orders, end-of-life issues, and technology utilization decisions have life-and-death consequences.

Several codes of ethics may help characterize the ethical behavior expected of clinical engineers and reflect the uniqueness of the clinical environment.

American College of Clinical Engineering code of ethics

The Principles of Ethics created by the ACCE in 2006 are as follows:

Preamble:

The following principles are established to aid individuals practicing engineering in health care to determine the propriety of their actions in relation to patients, healthcare personnel, students, clients, and employers. In their professional practices, they must incorporate, maintain, and promote the highest levels of ethical conduct.

Guidelines:

Principles of Ethics: published January 12, 2006

In the fulfillment of their duties, clinical engineers will:

- Hold paramount the safety, health, and welfare of the public.
- Improve the efficacy and safety of health care through the application of technology.
- Support the efficacy and safety of health care through the acquisition and exchange of information and experience with other engineers and managers.
- Manage healthcare technology programs effectively and resources responsibly.
- Accurately represent their level of responsibility, authority, experience, knowledge, and education, and perform services only in their area of competence.

- Maintain confidentiality of patient information as well as proprietary employer or client information, unless doing so would endanger public safety or violate any legal obligations.
- Not engage in any activities that are conflicts of interest or that provide the appearance of conflicts of interest and that can adversely affect their performance or impair their professional judgment.
- Conduct themselves honorably and legally in all their activities.

The uniqueness of the clinical setting is somewhat addressed in this code of ethics. The contributions to society as a whole as well as the vulnerability of the patient population are integrated subtly rather than explicitly.

Biomedical Engineering Society code of ethics

The BMES Code of Ethics, established in 2004, may also offer some insights into professional conduct.

BME is a learned profession that combines expertise and responsibilities in engineering, science, technology, and medicine. Since public health and welfare are paramount considerations in each of these areas, biomedical engineers must uphold those principles of ethical conduct embodied in this code in professional practice, research, patient care, and training. This code of ethics reflects voluntary standards of professional and personal practice recommended for biomedical engineers.

Biomedical Engineering Professional Obligations:

Biomedical engineers in the fulfillment of their PE duties will:

1. Use their knowledge, skills, and abilities to enhance the safety, health, and welfare of the public.
2. Strive by action, example, and influence to increase the competence, prestige, and honor of the BME profession.

Biomedical Engineering Healthcare Obligations:

Biomedical engineers involved in healthcare activities will:

1. Regard responsibility toward and rights of patients, including those of confidentiality and privacy, as their primary concern.
2. Consider the larger consequences of their work in regard to cost, availability, and delivery of healthcare.

The BMES Code of Conduct suggests that patient privacy is the primary focus, although one could argue that the clinical engineering profession demands that both safe and effective healthcare is the primary concern. The safest place for a patient to be is not in a hospital; there is no risk of malpractice, ineffective treatments, or privacy breeches. However, the safest and most private person who is not in a hospital may also die because a lack of medical care can cause death. Clinical engineers must serve as part of the healthcare team that balances the safety of treatment with the efficacy of medical care.

Future opportunities for clinical engineering

Several clinical engineering handbooks were written 20 to 30 years ago. In reviewing the authors' vision of the future, it is clear that the ability to predict the technological future is challenged by both imagination limitations and unpredicted innovation. In addition, a consistent theme was present in these handbooks: a wish for a time when clinical engineering as a profession is understood and valued. One author described the field of clinical engineering as a "constant struggle," and this seems true today. In reviewing the outdated references, however, the tremendous advances since publication are critical to help understand the future. For example, the square-wave defibrillator described in a 1980 text has long been updated to a far more clinically successful biphasic waveform. Transitioning all defibrillators to biphasic devices in the 1990s was no easy task, but the improved clinical outcomes have been dramatic.

Clinical engineers must be proactive rather than reactive as the future of medical technology emerges. The importance of continuing education serves as a tool to glimpse the evolution of healthcare. Medical care is shaped by several forces beyond technology including economics, societal shifts, and regulatory developments. For example, the debate over universal healthcare availability and models is not relegated to government officials. Understanding the impact of a diminished uninsured population utilizing clinical services in a particular facility profoundly affects a wide variety of technological support decisions at a local level. Another example is the movement of medical devices from hospitals directly into patient homes. Issues including proper use, education, maintenance, and management of devices that transmit data over non-hospital networks all affect support of these devices. Thus, clinical engineers need to focus on multiple, sometimes competing priorities. Keys to successful adaptation to change including tight connections to stakeholders and vigilance are crucial for continued clinical engineering success.

Abbreviations

AAMI	Association for the Advancement of Medical Instrumentation
ACCE	The American College of Clinical Engineering
AHA	The American Hospital Association
ARRA	The American Recovery and Reinvestment Act
ASHE	American Society for Health Care Engineering
BME	biomedical engineering
BMES	The Biomedical Engineering Society
BMETs	biomedical equipment technicians
CBET	Certified Biomedical Equipment Technician
CCE	Certified Clinical Engineer
CHTM	Certified Healthcare Technology Manager
DNR	do not resuscitate
HIPAA	Health Insurance Portability and Accountability Act
The HITECH Act; HITECH	Health Information Technology for Economic and Clinical Health
HTM	healthcare technology management
ISO	independent service organization
IT	information technology
NFPA	The National Fire Protection Association
NESCE	New England Society of Clinical Engineering
PE	professional engineer
VA	The Veterans Administration
WHO	The World Health Organization

References

About ACCE: Our Mission. (n.d.). Retrieved from <https://accenet.org/about/Pages/Mission.aspx>.

About ACCE: Clinical Engineer. (n.d.). Retrieved from <https://accenet.org/about/Pages/ClinicalEngineer.aspx>.

American College of Clinical Engineering. (n.d.). *Certification program*. Retrieved from <https://accenet.org/CECertification/Pages/Default.aspx>.

Davis, M. (1992). Codes of ethics, professions, and conflict of interest: A case study of an emerging profession, clinical engineering. *Professional Ethics*, *1*(2), 180–195.

Dyro, J. (2004). *Clinical engineering handbook*. Burlington, MA: Elsevier Academic Press.

FAQs about BME. (n.d.). Retrieved from Biomedical Engineering Society <https://www.bmes.org/content.asp?contentid = 140>.

National Academy of Engineering and Institute of Medicine. (2005). *Building a better delivery system: A new engineering/health care partnership*. Washington, DC: National Academies Press.

World Health Organization. (2017). *Human Resources for medical devices: The role of biomedical engineers*. Geneva: The World Health Organization.

CHAPTER 2

Healthcare technology basics

General types of medical technologies utilized in clinical settings

FDA definition and categorization

A medical device is defined in the Federal Food, Drug, and Cosmetic Act as "... an instrument, apparatus, implement, machine, contrivance, implant, in vitro reagent, or other similar or related article, including a component part, or accessory which is: recognized in the official National Formulary, or the United States Pharmacopoeia, or any supplement to them, intended for use in the diagnosis of disease or other conditions, or in the cure, mitigation, treatment, or prevention of disease, in man or other animals, or intended to affect the structure or any function of the body of man or other animals, and which does not achieve any of its primary intended purposes through chemical action within or on the body of man or other animals and which is not dependent upon being metabolized for the achievement of any of its primary intended purposes" (Title 21 Code of Federal Regulations [CFR], 2019).

The Food and Drug Administration (FDA) has established three regulatory classes based on the level of control necessary to ensure the safety and effectiveness of the device. Device classification depends on the intended use of the device and also on the indications for use. Device classifications determine the type of submission/application required by the FDA (2018).

1. Class I devices present the lowest level of risk for injury and are not intended to support or sustain life. Examples of Class I devices include elastic bandages, manual stethoscopes, examination gloves, and handheld surgical instruments.
2. Class II devices present a slightly higher level of risk for injury because they are more likely to come into contact with the patient. Examples include infusion pumps and powered wheelchairs.
3. Class III devices usually sustain or support life, are implanted, or present a potential unreasonable risk for illness or injury. Examples include pacemakers and automated external defibrillators.

The FDA definition also can be used to categorize medical devices into three areas: equipment used in the diagnosis, treatment, or prevention of a

Introduction to Clinical Engineering
DOI: https://doi.org/10.1016/B978-0-12-818103-4.00002-8

disease. These three broad categories can help identify the general types of medical technologies used in clinical settings in different ways.

Diagnostic devices gather physiologic or anatomic data to aid in the identification or treatment of illness, injury, or disease. Equipment used in the diagnosis of diseases can be simple or complex. Simple equipment includes items such as thermometers and otoscopes, which are used by clinicians for possible diagnosis of diseases such as infections. More complex technology can include physiologic monitors, which detect and record electrocardiogram rhythms to aid in the diagnosis of conditions such as atrial fibrillation. This category also includes many laboratory devices that may use a blood sample to determine blood counts or the presence (or absence) of a specific chemical marker, which aid in the diagnosis of disease states.

A large group of diagnostic devices are associated with imaging modalities that offer clinicians tools to visualize internal organs, bones, or other structures. Traditional x-ray units use radiation to provide simple views of bones, teeth, and lungs. More complex devices like magnetic resonance imaging (MRI) and ultrasound scanners offer greater insight into function, such as heart valve opening and closing, without the use of radiation.

Equipment used in the treatment of a disease is often termed *therapeutic* and may include items such as a defibrillator that uses an electric shock to synchronize the electrical activity of the heart or a hemodialysis machine that filters the blood to remove toxins. Ventilators are an important group of therapeutic technology, supporting vital respiratory activity. These devices treat the underlying condition or disease.

Equipment used in the prevention of disease may include an insulin pump that infuses a drug (insulin) to maintain a pre-prescribed concentration of the drug in the body or a left ventricular assist device (LVAD) used to take over a person's heart function as a bridge to transplant.

Table 2.1 shows the crosswalk of FDA classification of medical devices against the general categories defined in the Federal Food, Drug, and Cosmetic Act.

EQ89 standard definition

The Association for the Advancement of Medical Instrumentation (AAMI) in collaboration with American National Standards Institute (ANSI) have established a voluntary medical device standard that defines medical equipment as "medical devices that have been cleared by the FDA that are intended to be used for diagnostic, therapeutic, or monitoring care provided to a patient by a

Table 2.1 Crosswalk of FDA classification of medical devices against the general categories defined in the Federal Food, Drug, and Cosmetic Act.

FDA classifications of device	Federal Food, Drug, and Cosmetic Act general categories of medical devices		
	Diagnosis	**Treatment**	**Prevention**
Class I devices	Manual stethoscopes	Elastic bandages	Examination gloves
Class II devices	Physiologic monitors	Hemodialysis machines	Insulin pumps
Class III devices	HIV diagnostic tests	Defibrillators	Left ventricular assist devices

health care organization. Medical equipment includes devices such as monitoring equipment, life-support equipment, imaging equipment, laboratory equipment, mechanical equipment, transport equipment, as well as any other equipment supporting the care of a patient, whether or not it is in the immediate vicinity of a patient. In addition, this category includes other devices, such as computers, that support the care of a patient when in a health care organization, but are generally not specifically manufactured for use in a health care organization" (EQ89). This standard more broadly includes technology that is involved in patient care but does not fit into the FDA definitions.

Regulatory definition and categorization

Although the FDA's definition and categorization of medical devices sets the standard on how device manufacturers classify and determine the type of submission they complete to market their products, these classifications are not readily used in clinical engineering practice. Those departments managing medical equipment more closely follow categorizations and definitions provided by US healthcare regulatory bodies such as the Centers for Medicare and Medicaid Services (CMS) and the Joint Commission. These regulatory bodies govern hospitals and generally categorize risk based on harm to the patient or caregiver if the device fails.

High-risk (critical) equipment vs. non–high-risk (non-critical) equipment

High-risk equipment includes items for which there is a risk for serious injury or death to a patient or staff if the equipment fails. High-risk

Table 2.2 Crosswalk of FDA classification of medical devices against the Joint Commission classification of medical devices.

FDA classification of devices	The Joint Commission classification	
	Non–high-risk	**High-risk**
Class I devices	Manual stethoscopes	N/A
Class II devices	Physiologic monitors	Ventilators
Class III devices	N/A	Defibrillators, left ventricular assist devices

medical equipment encompasses life- support equipment, including items such as anesthesia machines and ventilators. The term *high-risk* is somewhat new to the Joint Commission medical equipment standards and is considered comparable to the CMS term *critical equipment*. The term *high-risk* is defined in the glossary of the Joint Commission standards (the Joint Commission, 2018).

The Joint Commission categorization of non–high-risk is akin to the CMS term *non-critical equipment*. Should non–high-risk equipment or systems fail, the likely result is limited to minor injury to patients or caregivers (National Fire Protection Association, 2018). Examples of non–high-risk (non-critical) equipment include examination tables, blood pressure devices, and physiologic monitors.

In general, these categories roughly align to the FDA classifications as shown in Table 2.2.

Devices throughout the healthcare system and relationship to patient care

Introduction

The delivery of treatment to improve the medical condition of a patient has evolved tremendously from the emergence of healthcare facilities in the 1800s. The tools and techniques of medical care have dramatically expanded. Beginning in the 1980s, technology expanded to assume a critical and impactful role as a component of the diagnosis and treatment of patients. The explosion of device types and device usage may have been driven by a risk-averse medical environment. Thus, clinicians utilized technology regularly as tools to avoid errors or the potential perception of negligence. As a result, healthcare staff have described a shift in role from the "care of patients" to the "care of technology." This framework is critical to clinical engineers as they understand the healthcare system and medical practice.

Medical care is generally divided into two categories: acute and chronic. Acute care is associated with illness or injury, requiring treatment for a period with a goal of recovery or wellness. In contrast, chronic care is associated with long-term or permanent conditions that require medical care for an extended period. In many cases, chronic care is not focused on recovery but instead on support of the quality of life of the patient. Technology is closely integrated into both types of care, but the majority of devices are involved in acute treatment.

Patients within a healthcare system are categorized as either inpatient or outpatient. A general misunderstanding exists surrounding the definition of these two categories related to the assumption that an overnight stay at a hospital or healthcare facility designates the encounter an "inpatient" event. Instead, a patient receives inpatient status when he or she has been formally "admitted" to the hospital with a physician's order. In this way, status is administratively assigned by the medical staff. Patients have outpatient status when receiving services in the emergency department, during observation, when undergoing surgical treatment without admission (outpatient surgery), and during laboratory tests, x-ray imaging, or other services that are performed without a physician's order for admission (US Centers for Medicare and Medicaid Services, 2018).

Examples to help identify the differences between inpatient and outpatient status are described in the following scenarios.

Scenario 1: Outpatient surgery/procedures

A patient had a biopsy scheduled in the endoscopy department. The patient arrived for the procedure, and the test produced some excessive bleeding. The patient remained in the recovery area for several extra hours for observation. However, at no point was there a physician order to admit the patient, and so this entire treatment is considered an outpatient event.

Scenario 2: Clinic visit (stress test) to emergency department to surgery

A patient had a clinic visit (outpatient) for a stress test at his physician's office. During the stress test, the technician observed issues related to the patient's heart. The technician conferred with the physician who, based on the symptoms, recommended that the patient be transferred to the emergency department for additional tests to rule out a myocardial infarction (MI).

When the patient arrived at the emergency department, he completed an additional non-invasive imaging study and laboratory tests to diagnose an MI. At this point, the patient remained in outpatient status because no physician had ordered admission to the hospital. The test results were inconclusive. The physician decided to place the patient in "observation" status that allowed a designated period (23 hours to several days, depending on the institution) to determine whether to admit the patient. The patient remained in the hospital during observation and was placed in a typical hospital room on a patient care floor outside the emergency department.

After overnight observation, additional test results showed that the patient had a heart condition and that surgery was warranted. At this point, the physician wrote an admission order, and the status of the patient changed to inpatient. The patient was scheduled for inpatient surgery and continued his stay in the hospital.

The definition of inpatient versus outpatient is important for many financial reasons associated with health insurance reimbursement. In addition, the types and complexity of medical equipment vary across these different locations.

Outpatient environments

Clinics

Short treatment duration is a hallmark of this environment of patient care. Clinic environments can range in complexity from general primary care locations to complex procedural care locations. General primary care clinics utilize basic, non-critical equipment such as patient scales, thermometers, examination tables, otoscopes/ophthalmoscopes (lighted devices to examine ears or eyes), and sphygmomanometers (a blood pressure cuff with a pressure gauge). Specialty facilities such as orthopedic clinics may have other non-critical medical equipment such as cast saws and basic x-ray imaging devices. Physical medicine and rehabilitation clinics may utilize exercise equipment (e.g., bicycles and treadmills), transcutaneous nerve stimulators, water baths, and other specialty non-critical equipment. Urgent care locations are another type of specialty location that manages non—life-threatening illness and injury, often needed when primary care physicians are not available. Urgent care facilities feature a broader range of technology, including both primary care location devices, and they may utilize additional diagnostic devices such as laboratory equipment to test blood and urine specimens. Some urgent care facilities may also have imaging devices such as x-ray machines and simple ultrasound scanners. All equipment in urgent care locations is considered non-critical except for emergency response devices such as defibrillators.

Clinics may focus on procedures or tests. In these areas, devices are based on the type of activities performed. For example, locations that perform stress tests will have treadmills and physiologic monitors. Endoscopy clinics use specialized scope equipment such as various gastrointestinal scopes or bronchoscopes, to view various internal parts of the body through a camera within the scope. Most of this equipment is categorized as non-critical, however there may be a few pieces of critical equipment associated with emergency response such as a defibrillator. In addition, anesthesia machines may be utilized when sedation is provided during various clinical activities.

Dialysis clinics feature kidney dialysis machines that replace the function of the kidneys. A semipermeable membrane filters the patient's blood to remove waste products normally separated by the kidney. Dialysis devices require a wide variety of mechanical and electrical components to manage the fluid filtration.

A note about emergency response technology: Cardiac fibrillation is a life-threatening medical condition that generally requires rapid intervention to prevent patient death. Delivery of an electrical signal through the chest to the heart can reverse fibrillation. A defibrillator is a cardiac care device that delivers energy within specific energy levels and utilizes a therapeutic waveform to permit the conduction system in the heart to resume the periodic stimulation that promotes successful movement of blood through the heart. Defibrillators are available for use in almost all patient care environments and often are a component of an emergency response cart, sometimes called a *crash cart* or *code blue cart.* This collection of medications and tools needed for resuscitation may require frequent performance testing and assessment to ensure proper delivery of life-saving interventions when needed. Also very common, an automated external defibrillator (AED) requires no medical training to utilize and is stored in many buildings similar to fire extinguishers, as well as public venues such as schools, theaters, and airplanes. An important feature of AED performance ensures that patients whose cardiac activity is not fibrillating do not receive a dose of energy. The AED detects the electrocardiogram (ECG) and ensures that the right amount of energy is delivered at the right time, should it be warranted.

Emergency department

Emergency departments utilize equipment that ranges from basic triage equipment (e.g., stretchers, thermometers, sphygmomanometers, suction machines, ECG machines, intravenous [IV] pumps, physiologic monitors) to highly complex equipment used to treat traumatic injuries such as car accident or gunshot victims. Trauma equipment includes treatment bays complete with a wide range of dedicated equipment including x-ray machines, ultrasound scanners, warming devices, crash carts for emergency resuscitation, ventilators, intubation devices, and incubators. Emergency departments utilize equipment categorized as high-risk (serious injury or death to a patient or staff if the equipment fails) such as ventilators and defibrillators.

A note about patient monitoring: technology enables medical staff to continuously monitor various physiologic parameters that can offer information about the health of a patient. Tracking patient health, especially during a visit to the emergency department, can be critical to ensuring that undiagnosed or changing illness or injury is detected and addressed. Tracking the electrical activity of the heart using a simple ECG monitor can offer a wealth of information regarding patient health. Three-electrode ECG monitors yield three physiologic views of the heart. Five-electrode ECG monitors produce six physiologic views of the heart. A 12-lead ECG monitor uses 10 electrodes but offers 12 views of the heart. However, caution should be used here as many engineers translate the word *lead* to mean an electrical connection. Physicians use the term *lead* to mean a physiologic view of the heart. Twelve-lead ECG devices are typically used as a diagnostic tool rather than an ECG monitor.

Even simpler monitoring technology utilizes light sensors that can transcutaneously detect the oxygenation of a patient's blood, typically in the extremities. Pulse oxygenation devices, sometimes called *pulse ox* or *SpO_2*, detect a rise and fall in blood oxygenation at each heart contraction. As a result, heart rate is also detected. Because pulse oximetry devices are low in cost, simple to use, and require only one connection to a finger (rather than the sticky ECG pads placed on the torso needed to detect electrical activity of the heart), this technology is extremely common. Please note that pulse oximetry devices do not detect respiration rate, although respiratory effectiveness is related to the oxygenation in the blood.

One additional monitor type is common in many clinical care areas. Automated blood pressure devices detect tiny blood pressure fluctuations (oscillations) in the patient's arteries, deflating a blood pressure cuff to determine the systolic and diastolic pressures. Often integrated with other vital sign tools, such as a thermometer and pulse ox sensor, the monitor enables

(*Continued*)

(Continued)
medical staff with limited technical training to gather periodic information. As opposed to the use of a stethoscope and sphygmomanometer, the automated device does not require careful listening skills to accurately measure vital signs.

Observation units

Observation units vary based on the needs of the patients and institutions. Some units serve less serious illnesses or injuries and, as a result, require less complex technology including hospital beds, thermometers, patient scales, and blood pressure monitors. In contrast, many observation units are hospital-based locations that have standard equipment found in most patient care areas including hospital beds or stretchers, air/oxygen regulators and blenders, thermometers, blood pressure monitors, emergency response carts with defibrillators, IV pumps, and physiologic monitors.

A note about the delivery of medication and fluids through the human venous system: Delivery of medications to patients can be oral (pills), by injection, through the skin (transdermal), or through the circulatory system (IV infusion). Most patient care areas utilize medical devices to assist in the administration of drugs into the circulatory system through the use of mechanical/electrical pumps. Treating simple dehydration through fluid delivery into the venous system in a variety of outpatient settings commonly utilizes IV pumps. In most hospitals, the highest number of devices owned is associated with IV pumps. The use of technology ensures accurate dosing, and includes automated alarms when occlusion is detected and when administration is complete. Pumps have a variety of specialized features, including patient-controlled analgesia (PCA) pumps that deliver pain medication. The devices allow delivery of the medication (within certain parameters) when the patient pushes a button. Controls are lockable and inaccessible to the patient. Many types of pumps are termed *smart pumps* that feature drug libraries and other safeguards to assess dosages and drug interactions. The goal of smart pumps is to reduce medication errors.

Home health

Medical equipment has been migrating out of the hospital and clinic environment into the patient's home. Healthcare is provided to patients in their home at lower cost and with better clinical outcomes. The utilization of medical technology in this care is expanding to include complex equipment, including high-risk devices such as ventilators. Physicians may prescribe basic equipment such as patient scales, blood pressure or pulse oximetry monitors, glucose meters, respiratory support devices, and nebulizers. High-risk equipment may also be used in the home, such as LVADs for heart failure patients, phototherapy equipment for infants, or ventilators for patients who cannot breathe on their own. Equipment may require remote or in-person technical support to ensure proper functioning.

Home healthcare may offer improved clinical outcomes, but technology use can offer a variety of challenges. Caregivers must manage devices designed for use in the clinical setting by medical staff. Family members may lack a general understanding of physical properties such as electricity and pneumatics and may face literacy hurdles. In addition, the home environment may feature an unreliable electrical supply, have varying humidity or temperature levels, or be unsanitary. Clinical staff who manage in-home patients who utilize complex technology must prepare caregivers for malfunctions or other emergencies. Clinical engineers must plan for performance assurance plans and preventive maintenance during long-term use.

Telehealth

Hospitals and health systems utilize telehealth solutions to deliver care to patient populations in rural areas to minimize travel requirements, or for patients in areas where specialty care is not readily available. In these outpatient settings, monitoring and diagnostic interventions are conducted while the patient is in his or her home. Cardiac or fetal monitors are examples of devices provided to patients to gather physiologic data for medical interpretation. Remote monitoring of implantable defibrillators or pacemakers are other applications of telehealth. In this environment, care often requires a secure and reliable Internet connection and may utilize video cameras.

Inpatient patient care areas

Operating room

Surgical procedures of many types are performed in this specialized care area. Operating rooms feature a very unique and highly controlled environment that contains a great deal of equipment. Operating rooms can be general or highly specialized based on the types of surgeries performed in the suite. General operating room equipment includes an operating room table with specialized surgical lights, which are often controlled with foot pedals. Anesthesia machines deliver specialized medications to induce paralysis, unconsciousness, and pain relief, often with several gases including nitrous oxide, sevoflurane, desflurane, isoflurane, and halothane. Paralysis results in the need for anesthesia machines to utilize a ventilator to deliver breaths to the patient. In addition, capnography is used to determine the amount of expired carbon dioxide, indicating the effectiveness of gas exchange. Anesthesia machines also utilize absorbers to ensure that anesthetic gases are removed from the ventilator tubing.

For many years, determining the depth of anesthesia experienced by patients was established as an art rather than a science. However, monitoring the depth of anesthesia utilizing electroencephalography (EEG) to detect electrical activity in the brain can be used to quantify a patient's sedation on a scale of 0 to 100. A small strip of EEG electrodes attached to the patient's forehead can be analyzed to guide the delivery of anesthetics.

Electrical waveforms are used in surgical procedures to cut the patient as well as to coagulate blood vessels. The technology is termed *electrosurgical units (ESUs)* and generates signals in the RF range of 300 to 3000 kHz. A plume (smoke) is generated during the use of this device, and, as a result, smoke evacuators are needed during most surgical cases. ESUs feature two modes, a monopolar mode and a bipolar mode. The monopolar mode uses a singular probe (pencil) and a return electrode placed elsewhere on the patient, and the bipolar mode features a tweezer-like instrument that passes electrical current across a very short area.

Light amplification by stimulated emission of radiation (LASER) is utilized in surgical cases because the light wavelengths can be refined to selectively affect tissue, such as tumors, tattoos, and skin resurfacing. Many surgical cases use a common carbon dioxide LASER for cutting and coagulation. Specialized LASERs include Nd:YAG (neodymium yttrium aluminum garnet) and argon.

Minimally invasive surgery procedures utilize rigid or flexible scopes that feature fiber-optic light sources with a video camera to view inside very small spaces in the body, including the digestive system, abdominal cavity, and joints. During these surgical cases, tools are inserted into the body through tiny openings, guided by the video camera, to make anatomic changes. Types of scopes are associated with the part of the body to be entered or viewed and include laparoscopes, endoscopes, gastroscopes, proctoscopes, duodenoscopes, bronchoscopes, and cystoscopes. When the abdominal cavity is explored using laparoscopic surgery, an insufflator is used to create a space between the abdominal wall and the organs. The device adds carbon dioxide to the cavity.

Surgical microscopes offer medical staff the opportunity to view small structures. Surgical microscopes are typically binocular and purely mechanical, often on booms for better positioning around the patient.

Computers assist in surgical procedures in a variety of ways. Surgical navigation allows the integration of imaging tools to be utilized in real time during cases. Surgical robotics utilize mechanical devices, guided by surgeons, to manipulate surgical tools. Joysticks and other controls permit highly refined actions. These tools are highly complex and special-purpose technologies that can support specific surgical procedures when high precision is needed.

Maintaining patient temperature during surgical procedures utilizes several technologies. Blanket warmers circulate warm air through plastic blankets. Electrical warmers can bring blood or IV fluids to body temperature. Radiant warmers are often mounted to the ceiling.

Procedural areas (cardiac catheterization, endoscopy)

Procedural areas are similar to operating rooms, but they are generally more focused on the test/procedure being undertaken. In addition to the general equipment found in an operating room (table, lights/booms, anesthesia machine, crash cart), equipment associated with the specific procedure is included in these areas. In a cardiac catheterization laboratory, minimally invasive tests and procedures are performed to diagnose and treat cardiovascular disease. Specialized equipment required includes single- or multi-plane imaging equipment, patient cooling devices, intravascular ultrasound scanners, and injectors. Endoscopy suites provide non-surgical procedures to view the digestive tract. Specialty equipment includes endoscopes, scope washers, ultrasound scanners, and video systems. Other procedural areas may also include diagnostic/interventional radiology suites and radiation oncology procedure areas, featuring specialized imaging technology.

Pre-procedure and post-procedure rooms/bays

Pre-procedure rooms/bays are generally used to prepare patients before procedures and surgery. These spaces hold patients temporarily and generally have high patient turnover and throughput. Following surgery or a procedure, patients are monitored in a recovery area, often categorized as a post-anesthesia care unit intensive care area. Equipment in these areas includes stretchers/beds, physiologic monitors, blanket warmers, and IV pumps.

Intensive care unit

High-acuity patients who are very ill or in serious condition require a significant amount of medical care and support and are generally admitted to intensive care units (ICUs) that utilize a large amount of technology. Hospitals may have ICUs that are specialized to certain patient populations such as cardiology and neurology. High-end equipment in cardiac ICUs may include advanced life-support equipment such as LVADs or extracorporeal membrane oxygenation (ECMO) equipment, which is used for patients with life-threatening heart and/or lung problems. ECMO oxygenates blood and removes carbon dioxide from the patient's body in a similar way to heart-lung bypass machines utilized for cardiac surgery. Cardiac units also include intra-aortic balloon pumps that assist the heart in generating pumping pressure through a catheter inserted into the descending aorta, which allows a patient's heart to rest and heal.

Ventilators, also called *respirators*, support the respiratory activity of patients. The devices are complex and can push air into the patient's lungs at a preset rate; support patient efforts to breathe on their own; deliver air if the patient goes too long without taking a breath; or finish when the patient initiates a breath. One type of ventilator offers high-frequency ventilation, delivering tiny volumes at incredibly fast rates of 500 breaths per minute and beyond.

In neurology ICUs, equipment such as hypothermia devices, intracranial pressure monitors, and ventilators may be used. Other general ICUs may also have specialty beds, dialysis machines, ventilators, IV pumps, feeding pumps, point-of-care laboratory analyzers, wound vacs, and high-end physiologic monitors that examine additional invasive pressure measurements. Some ICUs may also contain equipment booms and lifts to move patients.

Intermediate care unit/continuing care unit

When patients need a higher level of care than in a typical patient care unit but do not require the technology or medical staff attention of an ICU, they are often placed in an intermediate care unit (IMC) or continuing care unit, sometimes called a *step-down unit*. These units have some critical equipment such as ventilators and dialysis machines, but they generally do not contain the myriad of specialty ICU equipment. IMCs usually have physiologic monitors in each room. Telemetry units allow patients to be ambulatory while still recording cardiac activity with wireless transmission. All of the general equipment in the acute care areas is also present in an IMC.

Acute care—common patient care rooms

Patients in need of a lower level of care are usually placed in the acute care unit, often categorized by the type of patient including medical, surgical, labor and delivery, oncology, and neurology. These units generally have vital signs devices for periodic data collection instead of continuous physiologic monitors. These portable devices episodically measure heart rate, blood pressure, and pulse oximetry. Other equipment in acute care areas may include sequential compression devices to minimize blood clots in patients' legs, IV pumps, and patient beds.

In most hospitals, the highest number of devices owned is associated with IV pumps. The use of this technology ensures accurate dosing by featuring automated alarms that signal when occlusion is detected and when administration is complete. Pumps have a variety of specialized features, including PCA pumps that allow patients to self-administer pain medication. Syringe pumps are used to deliver a very small amount of medications, mechanically moving the piston of a syringe to send medication into IV tubing. In contrast with other pumps, feeding pumps can deliver nutrition into the digestive system. Termed *enteral nutrition*, delivering calories into the digestive system is contrasted with *parenteral nutrition*, which utilizes the circulatory system and IV pumps.

Labor and delivery

The majority of births in the United States occur in medical facilities. Many hospitals dedicate significant resources to creating a welcoming and warm environment that patients will select for the birth of their baby. During labor and childbirth, several specialized devices are utilized to monitor both the patient and fetus. Tracking the ECG activity of unborn babies is critical in the assessment of the stress of the delivery process. If electrical activity were tracked, the signals of both the mother and baby would be combined and difficult to interpret. Therefore, fetal heart monitors utilize transcutaneous Doppler measurement techniques to detect the movement of the heart. In some cases, when the fetal head is accessible, a direct scalp electrode can be used to measure ECG electronically. In addition, contractions can be monitored with the use of a transducer placed on the skin. Some fetal monitors use a telemetry technique that can wirelessly transmit fetal heart rate and contraction strength from the sensors. This permits patients to ambulate during labor.

During recovery for the baby and mother in a labor and delivery unit, many hospitals feature infant abduction systems that track newborns, preventing unintended departure from the patient care area. Radio frequency identification tags, worn by the baby, and in some models, the mother, can also be used to avoid infant switching, in which a baby is sent home with the wrong parents. Generally, the tags sound an alarm if removed from the baby.

Pediatric and neonatal areas

Areas in the hospital that cater to pediatric and neonatal patient populations are organized into acute care, intermediate care, and intensive care. In addition to the equipment utilized in adult care areas, pediatric and neonatal areas may also have specialty items. Newborn babies within a neonatal intensive care unit (NICU) utilize incubators to support their growth and development. Incubators, also called *isolettes*, feature a small bed surrounded by Plexiglas. This environment can be controlled for temperature and humidity. Small doors allow access to the patient while minimizing changes to the patient care area. Radiant warmers are also used to treat infants in the NICU, featuring open beds that have a radiant heater and bright lights above the uncovered patient bed. A sensor placed on the baby controls the temperature provided by the heater above.

Neonatal patients may have a common medical condition termed *hyperbilirubinemia* (commonly called *jaundice*), which occurs when the liver struggles to manage red blood cells. To treat hyperbilirubinemia, patients are exposed to a specific wavelength of light (between 425 and 475 nm) that transcutaneously destroys the toxic compounds in the blood. This wavelength is in the blue region, which can produce undesirable side effects in the medical staff and can mask coloring changes in patients. As a result, fiber-optic phototherapy blankets or broad-spectrum light sources are used.

Pediatric intensive care units (PICUs) include specialty equipment like specialty cribs and oxygen tents. Pediatric units may also feature play areas including gaming consoles and other play technologies. Other devices in

NICU and PICU areas include scales and length boards, feeding pumps, blood/solution warmers, breast pumps, and milk warmers.

Other equipment used across the hospital

Headwalls are generally found in all units at the head of the bed and include standard equipment such as vacuum/suction regulators, air/oxygen flow meters, emergency power electrical outlets, air/oxygen blenders, and nurse call systems. Nurse call systems have evolved from the simple bedside buttons that turned on a light in the hallway to signal a request for a nurse to respond. Now nurse call systems are highly integrated with paging and other communication systems to allow conversation between the patient and caregivers before a staff member visits the bedside. Some facilities track response times utilizing hospital computer networks.

The beds in which patients rest are often quite complex, offering substantial capabilities and monitoring tools. Mattresses can move to improve circulation, feature air holes to continuously expose skin to air, and change shape to help patients sit and stand. Beds can monitor a patient's weight or sound an alarm if the patient leaves the bed. Orthopedic beds feature frames and lifts to promote bone healing. Bariatric beds can manage patients of extreme size, from 600 to 1000 pounds. Birthing beds feature removable parts and unique additions, such as bars, to support patients during labor and delivery.

Other specialty areas

Imaging

Imaging modalities are often categorized by the method in which images are generated: ultrasound, radiation such as x-rays, and MRI. Ultrasound images are created using sound waves and are able to gather real-time images of both anatomy as well as function (heart valve opening and closing). Ultrasound devices can be very small (the size of a tablet computer) and are generally portable. Although bone and air block ultrasonic waves, soft tissue areas such as the abdomen are excellent applications for ultrasound imaging.

Images generated by radiation include x-ray machines that are composed of a single x-ray source producing two-dimensional images. Fluoroscopy units (C-arms) offer real-time moving images generated by x-ray exposure. Angiography is a common application of fluoroscopy, imaging blood flow in vessels. Dual-energy x-ray absorptiometry (DEXA) scanners determine bone mineral density utilizing x-ray radiation. In addition, computed tomography (CT) scanners feature greater image clarity with the use of multiple x-ray sources and detectors. Mammography is an x-ray imaging technique that is specialized for breast tissue. Positron emission tomography (PET) scanners utilize radiation associated with altered glucose molecules to image glucose uptake, which is critical in cancer detection. PET scanners are often combined with CT scanners to provide anatomic and physiologic images at one time.

MRI creates images related to the exposure of hydrogen atoms to magnetic fields and radio waves. MRI images are profoundly useful as they depict physiology as well as anatomy because the hydrogen atoms are associated with body chemistry rather than just anatomic structures. Because of the strong magnetic fields needed for MRI imaging, equipment used to monitor or support a patient during an imaging procedure must be specialized and nonmagnetic so it can be present in MRI room during a scan.

Nuclear medicine

Nuclear medicine imaging provides patients with small amounts of radioactive materials to image very specific areas of the body. Unlike other imaging techniques, nuclear medicine imaging often uses gamma radiation. For example, radioactive iodine is utilized to image the thyroid gland by detecting the radiation using a gamma camera. The gamma radiation is designed to dissipate quickly from the patient. Nuclear medicine technology is also utilized to provide treatments to patients, often to address cancerous tumors, by delivering well-defined radiation to specific parts of the body.

Clinical laboratory

The clinical laboratory of a hospital utilizes samples of fluids or tissues from patients to identify evidence of disease or medical conditions. The space is organized into divisions such as anatomic pathology, clinical chemistry, hematology, genetics, microbiology, phlebotomy, and the blood bank. Some hospitals also have a reproductive biology testing division or a blood donor center that may or may not fall under the laboratory. Each section of the laboratory has specialized equipment and analyzers to conduct tests on blood and other specimens. Technology includes general equipment like microscopes, centrifuges, slide strainers, heaters, incubators, shakers, and tissue preparation devices. Specialty equipment can include apheresis machines, chemistry analyzers, hematology analyzers, electron microscopes, cell counters, and automated specimen processing lines. Many devices are highly complex and automated, processing samples, adding reagents, and making measurements in complex ways. Most laboratory instruments interface with a laboratory information system that receives the test results and sends them to the patient's medical record.

Pharmacy

Pharmacies generally prepare drugs for patient populations. There is little equipment in these areas except for PCA pumps that may be prescribed to a patient for pain. Larger pharmacies may also have technologies such as IV mixing robots, which may require support from the clinical engineering department. Many pharmacies deliver medications to the patient care areas through medication dispensing systems that limit the access of clinicians to drugs through computerized control of locked drawers and bins.

Sterile processing

The sterile processing department, sometimes called *central sterile services* or *central supply*, is responsible for sterilization or cleaning of reusable medical devices and consumables. In this department, various types of scope washers, sterilizers, and autoclaves are used to clean and sterilize the devices. Sterilizers may use high or low pressure, high temperature, steam, or ethylene oxide to destroy germs and to clean equipment and consumables.

Technology support life cycle

Introduction

Clinical engineers support technology throughout the equipment life cycle. This life cycle starts in the pre-acquisition phase, when hospitals and clinics or medical staff identify a technology need. Ideally, a requirements team/group will be established, connecting key stakeholders for a multifaceted decision process. Most of the time, the first step is to complete a pre-procurement evaluation or assessment to determine which specific devices will meet their needs. Once a selection is completed, the device implementation is planned. This includes the purchase of the device, as well as the tasks needed to install, integrate, and train staff on the use and support of the device. After the installation and training, the clinical engineering department supports the device on a day-to-day basis with preventive and reactive maintenance tasks. The clinical engineering staff also manages recalls, alerts, and incident investigations when

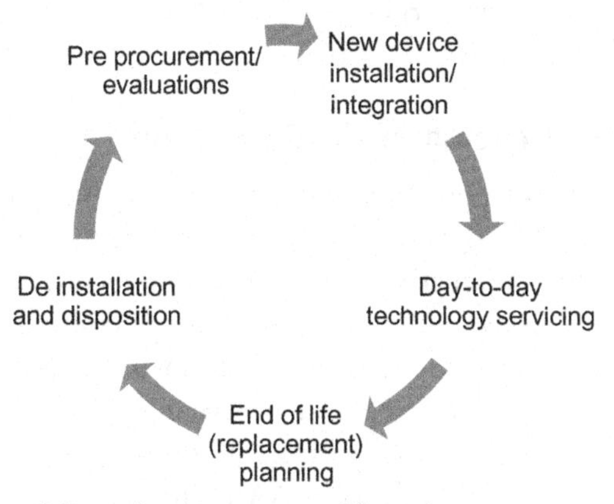

Figure 2.1 Phases of the technology support life cycle.

a malfunction is observed. When clinicians or clinical engineers determine that the end of a device's useful life has been reached (often as a result of high support costs or evolving technology), the clinical engineering department then plans for replacement or removal of the device as well as de-installation and disposition of the equipment. Fig. 2.1 shows the phases of the technology support life cycle.

Medical device inventory

The creation and maintenance of an accurate inventory of medical technology is critical to institutions to provide support over the equipment life cycle. The Joint Commission also requires an accurate inventory for all medical equipment. Almost all departments that support healthcare technology utilize a computerized maintenance management system (CMMS). Each device in the inventory should have a unique identifying tag as determined by the institution, often a sticker or metal plate with a number. Software to collect and house device serial numbers, model numbers, and other information associated with the specific asset tag is available from a variety of vendors. In addition, the software can track work order requests, track equipment repair histories, track repair part needs, and maintain detailed histories associated with individual devices. The use of a computerized tool can promote data analytics that can yield aggregated information for a particular device type or category. Broader analytics associated with multi-site information can be extremely useful in the prediction of repair needs or other interventions to ensure proper technology performance and device availability.

Pre-procurement evaluation and selection

The purpose of the pre-procurement evaluation is to compare and contrast the available FDA-approved device brands and types that are currently available on the market to meet defined clinical needs, system specifications, and fiscal constraints. Institutional culture may drive the device selection process, where some clinicians make independent procurement decisions, in contrast with other sites, where a requirements team composed of clinical engineers, medical staff, and other key stakeholders drives a multifaceted process. AAMI has produced an acquisition guide to help clinical engineers share best practices associated with device procurement.

Request for Information

Most hospitals use a Request for Information (RFI) process to gather information from vendors on available products. This process begins by gathering performance requirements and feature expectations from clinical users (e.g., physicians, nurses, respiratory therapists) and those who support the technology (e.g., clinical engineering, facilities, information technology, purchasing). These requirements may relate to device functionality and usability; physical size and weight of the device; integration with existing models, brands, and systems; specifics on how to clean and maintain the device; information on software/server requirements; cybersecurity information history of recalls/alerts; required disposable purchase contracts; and any other requirements. The supply chain/materials management department will help gather all of these requirements and submit them in a formal document to vendors who have products that meet the specifications. Vendors respond to the requirements and provide more information on the details surrounding their products. Usually the team that generates the requirements then reviews the responses to determine which products/vendors meet the minimum requirements.

Request for Proposal

Once the field of vendors/products has been narrowed, hospitals/health systems then perform a Request for Proposal (RFP) process, which not only includes requirements from the RFI, but asks for pricing and cost details, installation timelines and requirements, and specific legal requirements such as business associate agreements or cyber-language review. Proposals from the vendors are reviewed by the requirements team. A best practice utilizes on-site demonstrations of the equipment or loaners for pre-purchase trials. Reference calls to other customers may be completed.

Selection and acquisition

During this stage of procurement, the vendor is selected based on a variety of factors and value weighting. For example, recommendations from physicians may have more influence in the vendor/model selection than the choice made by other stakeholders. Once a vendor is selected based on functionality, support, and total cost of ownership, the supply chain team then goes to award the RFP. This consists of negotiating the specific terms and conditions related to the purchase, finalizing a scope of work detailing installation, adding any other support agreements, and finally signing the completed agreement or contract. A focus of many acquisition agreements is the availability of service manuals, passwords, specialized tools, and dongles required for device maintenance. Some clinical engineers incorporate National Fire Protection Association (NFPA) 99 healthcare facilities code 10.5.3.1.2 "Service manuals, instructions, and procedures provided by the manufacturer shall be considered in the development of a program for maintenance of equipment" (National Fire Protection Association, 2018) as a tool within conditions-of-purchase documents to ensure receipt of the materials needed to service the device.

New device installation/integration

After negotiations are complete and a purchase order is initiated, the requirements team may be dissolved and an implementation team is assembled to complete the installation of the technology. Ideally, all components needed for the scope of work are included in the implementation of the new technology. This may include facilities, if work is needed related to the lighting, water, heat/cooling, or other facility-related support item. Information technology may be involved if the device has software, needs a server, is connected to the Internet, sends or receives data from any other system in the hospital, or is cloud-based. Representatives from the education department may be involved to ensure that clinical users are trained before equipment deployment. When devices are part of a system, complete testing should be completed and signed off by the users and all other stakeholders before going live with the system.

During this implementation, support documentation clearly delineating roles and responsibilities in the support of all aspects of the device should be developed. This documentation should include an overview of the system, specifics on installation (including but not limited to architecture drawings, configuration information, and server settings), and details related to user support. Documents should also include any maintenance agreement (if there is one), a life-cycle plan including proposed dates for software and hardware updates, details on staff training to use the device, and any technical training needed by clinical engineering, IT, or facilities to support the device. Device training from the vendor may include the acquisition of user and maintenance manuals, attendance at formal classes or online tutorials, and the creation of helpdesk workflows to troubleshoot challenges. This documentation should be shared and available to all stakeholders.

The new devices must receive an asset tracking tool of some type, such as a unique number affixed to the frame and then added to the inventory, typically through the CMMS. In addition, the new technology should be integrated into the institutional technology plan. A medical equipment management plan (MEMP) is a document required by accreditation agencies such as the Joint Commission. Ideally, a MEMP is a written plan to oversee all aspects of technology management. The document organizes the many facets of technology life cycle management including inventory, support, repair, and documentation. Newly acquired devices are integrated into the plan with appropriate documentation. For example, a MEMP can offer staff insights into appropriate steps for various situations including device failure, replacement/backup device availability, and requesting service for failed devices. Preventive maintenance frequencies, risk assessment, and data-driven analysis of service histories are also integrated into the MEMP.

Specialized tools, passwords, or test apparatus necessary for performance verification are acquired before the incoming inspections. Some vendors require technician training before providing specialized equipment. Many devices can be evaluated for proper performance utilizing commercially available simulators or test loads. These test devices must be researched and acquired before the shipment of new technology or devices.

Unboxing and assembly of new devices can consume significant amounts of manpower when multiple devices are purchased. An incoming inspection before the initial deployment of any new device is required and a best practice. Incoming inspection procedures focus on the performance assurance of all product features. Calibration and verification are also integrated into incoming inspection activities.

Operation: repair and maintenance processes

Once equipment/technology is installed/live, the recurring preventive and reactive maintenance tasks are completed over the life of the equipment. Both preventive and reactive maintenance are documented by the clinical engineering department in a file that contains the entire history of the device.

Preventive maintenance is a periodic practice and includes but is not limited to visual inspections, validation of performance, and calibrations to detect physical defects, or damage to devices caused by rough usage. These procedures are outlined in the manufacturer's manuals and should be completed according to the schedule identified by the manufacturer, generally a minimum of once per year. Preventive maintenance activities include inspections, functional checks, and measurements associated with the electrical safety of the device. Four currents are generally associated with risk to the user and/or patient. The currents include:

- Earth leakage current: the current that flows between the power supply of the device through the ground of the power cord;
- Touch leakage current: the current that flows between a person if he or she touches any part of the device (e.g., chassis) and electrical/earth ground;
- Patient leakage current: the current that flows between patient connections (like ECG leads) and electrical/earth ground;
- Patient auxiliary current: the current that flows between two separate patient connections (e.g., between ECG leads).

NFPA regulations define limits for these currents. Commercially available safety analyzers offer technicians automated tools to measure these currents when the device power cord and patient leads are connected to the safety analyzer. More information on this is presented in Chapter 4, Safety and systems safety.

Preventive maintenance activities can offer clinical engineers insight into user behaviors with devices. For example, if power cords are consistently damaged during preventive maintenance inspections, engineers may seek to determine how the power cords are damaged and how to mitigate this issue. Sometimes staff/user training can be implemented to reduce equipment damage. When informed by data analytics associated with repair histories, preventive maintenance activities and interventions can minimize equipment downtime by avoiding impending malfunctions.

While preventive maintenance activities and intervals are generally described by the device manufacturer, organizations are permitted to develop an alternative equipment management (AEM) plan. AEM programs help organizations manage manpower in the most efficient way, allowing technicians to focus on equipment that is a higher risk to patients if and when they fail, in ways that are specific to the particular needs of the clinicians and patient populations served. AEM plans adjust inspection and preventive maintenance activities and/or frequencies from the manufacturer's recommendations based on:

- records provided by the organization's contractors or vendors;
- information made public by nationally recognized sources (e.g., peer-reviewed journals, ANSI/AAMI); and
- records of the organization's experience over time.

The Joint Commission Environment of Care manual chapter 02.04.01 indicates that the strategies of an AEM program must not reduce the safety of equipment and must be based on accepted standards of practice, such as the American National Standards Institute/Association for the Advancement of Medical Instrumentation handbook ANSI/AAMI EQ56: 2013, Recommended Practice for a Medical Equipment Management Program (the Joint Commission, 2018).

The Joint Commission does indicate a few specific types of equipment that cannot be put on an AEM plan, but instead must follow the manufacturer's recommendations. These include:

- equipment subject to federal or state law or Medicare Conditions of Participation in which inspecting, testing, and maintaining must be in accordance with the manufacturers' recommendations, or otherwise establishes more stringent maintenance requirements;
- medical laser devices;
- imaging and radiologic equipment (whether used for diagnostic or therapeutic purposes); and
- new medical equipment with insufficient maintenance history to support the use of alternative maintenance strategies (the Joint Commission, 2018).

When determining if a piece of equipment can be put on an AEM program, the Joint Commission requires that a qualified individual(s) uses written criteria to support the determination of whether it is safe to permit medical equipment to be maintained in an alternate manner that includes:

- how the equipment is used, including the seriousness and prevalence of harm during normal use;

- likely consequences of equipment failure or malfunction, including seriousness of and prevalence of harm;
- availability of alternative or backup equipment in the event the equipment fails or malfunctions;
- incident history of identical or similar equipment (2 years of history is the general benchmark); and
- maintenance requirements of the equipment per the manufacturer (the Joint Commission, 2018).

These criteria must be clearly documented, and it is required that justification (data) for an AEM for a particular device be kept to justify the decision. Additionally, it is best practice to review the data periodically to ensure that the device and history continue to support the need for the device to be in an AEM program. Finally, the devices that are included in an AEM program must be identified in the inventory.

Fig. 2.2 shows an example from used to assess and document AEM. Note that the form includes a space for equipment category. Required categories include risk group, consequence of equipment failures, serious and prevalence scores, availability of alternative equipment, manufacturer's maintenance requirements, and AEM requirements.

As an example of an AEM in practice, the manufacturer of an otoscope may recommend that the light bulb be evaluated for performance every year. However, the institution may decide to place the otoscopes within a certain clinic area on a "run to fail" program. This would allow technicians to address the otoscope light sources only when reported as nonfunctional. The AEM would document the seriousness of failure as low as well as the prevalence of harm when the device fails as low. The AEM form would also document how clinicians could locate other otoscopes as backup equipment to promote continuity in patient care. The documented AEM would be implemented with a minimum of 2 years of device history proving that the "run to fail" program does not affect patient safety. Finally, the specific otoscopes put under the AEM program would be identified in the CMMS inventory system.

Reactive maintenance is the unscheduled repair of malfunctioning equipment. Repairs may require minor interventions or major part replacement. Institutions may spend large sums of money to ensure that technology is functional and available, as equipment downtime can substantially affect patient care and cause significant financial loss. For example, a patient may prepare for a procedure with fasting or other restrictions. When a device malfunction makes the procedure impossible, opportunities to reschedule

Clinical Engineering Alternative Maintenance Assessment Form

Location (s) of Device (s):			
Device Category:			
Device Model (s)			
Manufacturer (s):			
Risk Group:			
Seriousness Score:			
Prevalence of Harm:			
Likely Consequences of Equipment Failure or Malfunction:			
Availibility of alternative or back-up equipment in the event the equipment fails or Malfunctions:			
Incident History of Identical or Similar Equipment:			
Manufacturer's Maintenance Requirements of the Device:			
AEM Maintenance Requirements of the Device:			
Date Written:		Biennial Review Completed:	
Director's Signature:		Director's Signature:	
Date Updated in CMMS System:		Notes:	
Program Manager's Signature:			

Figure 2.2 Example form used to assess and document alternative equipment maintenance.

with a repeated preparation may not be possible. Repair activities may include but are not limited to troubleshooting the root cause of the issue, replacing the parts, cleaning the system, calibrating the system, documenting the work completed, and returning the device to clinical use.

As mentioned earlier, some devices are appropriate to assign to a maintenance strategy that relies on reactive actions only, utilizing activities to return devices to functionality. AAMI has prepared a standard, EQ89, *Guidance for the use of medical equipment maintenance strategies and procedures*, which can offer insight into best practices for maintenance and repair. Section 5 of this document specifically addresses repair strategies including the "corrective only maintenance" approach. Factors associated with this approach are listed in the standard and include age of the device, availability of spare devices, and impact on patient care. The use of best practices published by AAMI and other groups can be helpful in selecting the best practices of device support.

The availability of parts needed for repairs is a critical component in the life-cycle planning of equipment. Technician experience as well as repair histories can offer some insight into the parts inventory needed to support rapid return of technology to patient care areas. Balancing the quantity and types of components in the inventory with the predicted needs can be both an art as well as a science, informed by CMMS analytics. In addition, the identification of parts vendors and comparison of costs can be a highly time-consuming task. One source may be the least expensive for one technology type, while another vendor may be a better choice for a different device group. The cost of one part can vary by hundreds of dollars among vendors, demanding vigilance in the procurement process. In addition to unit cost, the possible impact of downtime on patient care is a factor in the availability of parts. Group buying power from a variety of collaborations can reduce these challenges in some cases.

Institutions generally use a mix of device support strategies for maintenance and repair as described here.

- *Complete in-house support using the technicians on site.* On-site technicians provide all device support including all repairs and preventive maintenance. This strategy may also be called *time and materials* as the costs for the parts are paid for out of the department's budget. Training for the technicians must be evaluated and managed so that a person with the technical skills needed for support/repair is available when a device fails.
- *In-house first review with outside technical support as needed.* This approach allows in-house technicians to troubleshoot repairs as far as they can before calling in a vendor when needed. This strategy may be

combined with service contracts, allowing flexibility in the type of education required.

- *Preventive maintenance–only service contract.* For equipment that fails infrequently, another strategy may be to contract with a vendor to complete preventive maintenance only. With this type of contract, any repairs are usually done on a time and materials basis.
- *Time and materials service contract.* A contract is negotiated that utilizes an outside vendor to complete services as needed. This type of contract usually negotiates costs per hour for outside labor and travel and possibly a discount on parts costs. Costs for service are based on the time spent completing the repair and the cost of the parts needed.
- *Service contract tied to supplies/consumables.* This approach obligates the institution to purchase disposables or reagents in exchange for device support/service. Careful planning is needed to avoid generating large amounts of unneeded disposables in order to have the device maintained.
- *Complete service contract support.* All service is provided by an outside vendor with all device support, *preventive* maintenance, repairs, and so on, performed for the specific negotiated cost.
- *Depot repair/support only.* Devices are shipped to an off-site facility for all support activities including preventive maintenance, calibration, and repair. This approach demands additional inventory in the hospital to ensure that clinicians have an adequate supply of devices when some are off-site.

A significant challenge for many technicians is the "no problem found" result of a repair alert/work order request. The communication between the user who reports an issue and the technician is critical to a successful diagnosis and repair. An unidentified problem can require extensive time to evaluate the performance of every possible device function, many times yielding no detectable defect. Imagine taking a car to a repair shop without identifying the issue that initiated the drop-off. The myriad of complex automotive systems could require hours of verification if no particular malfunction is described by the driver. In the clinical setting, wasted technician time as well as equipment downtime can be minimized with highly managed communication pathways and work order request procedures.

During the life of the system, manufacturers may also issue *alerts*, *recalls*, or field modifications to the device. These alerts/recalls may require updates of the hardware, software, or both to the device. The work may be done by the manufacturer or by the clinical engineering department. All alerts/recalls and actions taken should also be documented in the device history file.

Also over the life of the equipment, the device may be part of a *patient incident* or event. This is defined as any event/failure that may cause temporary or permanent harm to a patient or caregiver. Chapter 4, Safety and systems safety, provides more detail on clinical engineering's role in patient incidents/investigations, however all investigations should be documented as part of the device history file.

Device retirement planning

Proactive planning of replacement/retirement of equipment helps institutions plan inventory shifts, predict large expenditures, and allocate appropriate capital to ensure that clinicians have the necessary technology to provide continuity of care for patients.

The American Hospital Association generates a list of equipment lifetime estimates that are generally used by finance departments to assign depreciation to capital assets. The identified lifetime may also be used as a basis for planning the average life of equipment. Other data that are useful in planning when to replace equipment include the total cost to maintain the device over its life, the number of failures/quantity of downtime for the device, as well as new technology that has come into the market. Data-driven decisions based on historical repair records can drive lifetime predictions and technology planning.

De-installation and disposition

Once equipment is no longer repairable, usable, technologically appropriate, or slated for replacement, the final task clinical engineering completes is the de-installation and disposition of the devices. Larger systems, including complex imaging technologies, may need a third party to help de-install the equipment from the location because of its size or regulatory requirements around radiation-generating sources.

When de-installing equipment, clinical engineers must ensure that any proprietary information or patient information, including protected health information (PHI) or personally identifiable information is removed from the hard drives of the device to mitigate unintended release of records. Once the equipment is de-installed and PHI is removed, the device can either be disposed of, resold, or reused elsewhere in the healthcare system. Donations to foundations and other charitable organizations for use in developing countries are also common.

Abbreviations

AAMI Association for the Advancement of Medical Instrumentation
AED automated external defibrillator
AEM alternative equipment management
ANSI American National Standards Institute
CMMS computerized maintenance management system
CMS Centers for Medicare and Medicaid Services
CT computed tomography
ECG electrocardiogram
ECMO extracorporeal membrane oxygenation
EEG electroencephalogram
ESU electrosurgical unit
FDA Food and Drug Administration
ICU intensive care unit
IMC intermediate care unit
MEMP medical equipment management plan
MI myocardial infarction
MRI magnetic resonance imaging
NFPA National Fire Protection Association
NICU neonatal intensive care unit
PCA patient-controlled analgesia
PET positron emission tomography
PHI protected health information
PICU pediatric intensive care unit
RFI Request for Information
RFP Request for Proposal

References

Association for the Advancement of Medical Instrumentation. (2015). EQ89: *Guidance for the use of medical equipment maintenance strategies and procedures*. Retrieved from <https://www.aami.org/productspublications/ProductDetail.aspx?ItemNumber = 2421>.

The Joint Commission. (2018). *2018 Joint Commission comprehensive accreditation manual*. Oak Brook, IL, Joint Commission Resources.

National Fire Protection Association. (July 16, 2018). *NFPA 99-2012 code*. Retrieved from <https://downloads.nfpa.org/codes-and-standards/all-codes-and-standards/list-of-codes-and-standards/detail?code = 99>.

US Centers for Medicare and Medicaid Services. Medicare Inpatient vs Outpatient (July 16, 2018). Retrieved from <https://www.medicare.gov/what-medicare-covers/what-part-a-covers>.

US Food and Drug Administration. (July 17, 2018). *Classify your device*. Retrieved from <https://www.fda.gov/medicaldevices/deviceregulationandguidance/overview/classifyyourdevice/ucm2005371.htm>.

US Food and Drug Administration. (2019). *Title 21 code of federal regulations*. Retrieved from <https://www.accessdata.fda.gov/scripts/cdrh/cfdocs/cfcfr/cfrsearch.cfm>.

CHAPTER 3

Healthcare technology management

Systems engineering theory as applied to the clinical setting

The mini-system surrounding medical technology utilized in patient care is highly complex and tightly integrated into the overarching healthcare delivery system. As a result, this chapter begins with *systems thinking* as a vital tool to understand the use of technology. Systems thinking promotes the inclusion of all facets and components of a process or problem. Medical care through technology is not delivered by a black box in isolation that may possess needed technical capabilities. Instead, the device is part of a complex system of multiple stakeholders, including clinicians, support staff, technicians, and patients, with various needs woven into the continuum of care that can involve complex system disturbances such as operational errors. Thus, a clinical engineer must think holistically and systematically to better understand the interactions and demands of the components of the system surrounding medical technology. Essentially, to support a medical device in isolation is to undervalue the critical and dynamic interactions between people and the environment.

Systems engineering seeks to analyze, understand, and control complex activities to meet objectives. Simply stated, systems engineering principles seek to manage complexity. As a result, clinical engineers must try to characterize the stakeholders and their functions, needs, and interactions to understand the behavior of the system. In addition, systems engineering principles offer insight into methods to measure the performance of the system, optimizing deliverables and minimizing risk when possible. Modeling, including mathematical tools and critical path analysis, can generate insight promoting process improvement and, thus, improved patient care. The International Council on Systems Engineering (INCOSE) offers a wide variety of resources that support efforts to apply systems thinking to the healthcare environment.

Introduction to Clinical Engineering
DOI: https://doi.org/10.1016/B978-0-12-818103-4.00003-X

Two important reports, one issued in 2005 titled *Building a Better Delivery System: A New Engineering/Health Care Partnership* (National Academy of Engineering and Institute of Medicine, 2005) and a second published in 2014 titled *Better Health Care and Lower Costs: Accelerating Improvements through Systems Engineering* (President's Council of Advisors on Science and Technology, 2014) identify the importance of engineering in healthcare. Both reports are readily available online and may be useful to support improved understanding of clinical engineering.

An important and unique quality separates systems in healthcare from other complex systems. Various components and subsystems were never designed as systems, but they evolved over time without an overarching or purposeful understanding of the interactions of the components. Many thousands of tasks, workflows, and connections have evolved in the clinical setting and involve many stakeholders, with almost all of the activities woven into an unintended system without optimization or planning. Thus, the clinical engineer may be the ideal person to lead an analytical review of healthcare delivery through multiple systems analysis, reducing redundancy, and improving efficiency and reliability.

Components of the mini-system associated with medical technology include the device, the operator, the patient care area/facility, and the patient. As technology has evolved, the components have expanded to include the home as a facility, caregivers as operators, and the hospital network as part of the facility. Increasingly complex mini-system elements drive a systems approach that is thoughtfully examined. The integration of many mini-systems such as pharmacy and central sterile supply departments engages the analytical skills of engineers and encourages broad systems thinking.

The human-device interface and human factors

The US Food and Drug Administration (FDA) has defined human factors as follows:

> *The application of knowledge about human behavior, abilities, limitations, and other characteristics of medical device users to the design of medical devices including mechanical and software driven user interfaces, systems, tasks, user documentation, and user training to enhance and demonstrate safe and effective use. Human factors engineering and usability engineering can be considered to be synonymous (FDA, 2016).*

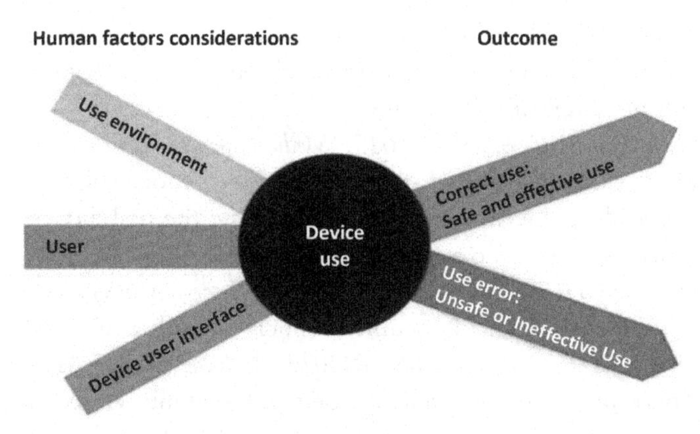

Figure 3.1 Human factors in device usage.

The importance of the interactions between a device and humans is critical for safe and effective technology use. Fig. 3.1 shows the relationships between a device and its use.

Clearly, the user is not the only consideration in a favorable outcome. The interface and the environment are additional challenges. Each consideration presents several components:

- *The user.* Clinicians are generally the users of many technologies, but their background, training, physical size, and understanding of the physiologic interface vary tremendously. In addition, as previously discussed, the home health environment may present caregivers as users who may have limited language proficiency, limited technical knowledge, or diminished sensory abilities.
- *The use environment.* While many treatment areas feature brightly lit clean spaces, devices may be used in darkened rooms or in loud spaces as well as in private homes. Design engineers may never have predicted conditions that can be associated with non-clinical sites, such as heat, humidity, or bugs.
- *The device-user interface.* Displays, commands, and alarms can be complex or simple to understand and manage. Knobs and other controls may be challenging for users with limited dexterity, difficult to understand, or difficult to clean. Instructions that may be well-designed for nurses could be too complex for in-home health aides or patients themselves.

Clinical engineers play a vital role in the identification of potential challenges associated with the components of safe device utilization, especially when new technology is deployed or existing technology is brought to a new environment. To guide some analysis and support safe device use, AAMI has created several standards and recommended practices associated with human factors including:

- *ANSI/AAMI/IEC 62366-1:2015, Medical devices—Part 1: Application of usability engineering to medical devices* specifies a process for a manufacturer to analyze, specify, develop, and evaluate the usability of a medical device as it relates to safety.
- *ANSI/ANSI HE75:2009(R)2013, Human factors engineering—Design of medical devices* is a detailed document associated with device design.
- *ANSI/AAMI/IEC 62366:2007(R)2013, Medical devices—Application of usability engineering to medical devices* defines a usability-engineering process that ensures the basic safety and essential performance of medical devices related to usability.
- *AAMI TIR49:2013, Design of training and instructional materials* for medical devices used in non-clinical environments.
- *AAMI TIR50:2014, Post-market surveillance of use error management.*

To clarify, many human factor elements are associated with device development and design, with good design optimizing the human interaction. Thus, technology arrives at a clinical site ready for implementation without the on-site clinical engineering input into usability and ergonomics. However, applying human factors analysis to the user-technology interface once technology is deployed in the clinical setting can promote safe and satisfactory usage. To aid in understanding the human-machine interface, the Human Factors and Ergonomics Society hosts an annual health care symposium featuring many workshops and sessions specifically aimed at post-implementation issues.

Computerized maintenance management systems

Computerized maintenance management systems (CMMSs) are the backbone to any clinical engineering department. These software systems are relational databases that manage maintenance activities (Cato & Mobley, 2002). These activities include but are not limited to the following:

1. Asset tracking
2. Work order management
   a. Preventative maintenance work orders
   b. Reactive maintenance work orders
   c. Other types of work orders
3. Labor tracking

4. Parts management
5. Contracts management
6. Reporting

Asset tracking

Asset-tracking activities include the management of assets that are the responsibility of the department. In most hospital institutions, the assets managed include all medical equipment, regardless of whether it is supported by the in-house clinical engineering team or outsourced to the original equipment manufacturer (OEM) or a third party for service.

Most hospitals tag their assets with a unique number that identifies the asset. This tag number along with the following information are populated into the database so that the asset can be identified:

- *Asset tag and/or property tag.* Some institutions use the finance property tag as their asset tag, and others have both a unique clinical engineering asset tag and a separate finance property tag.
- *Make, model, and manufacturer.* This details what the device is and who manufacturers it (not who you purchased it from).
- *Serial number, system ID, and/or unique device identifier (UDI).* The serial number is an alphanumeric number provided by the manufacturer that identifies the asset uniquely. The system ID is another alphanumeric number generally assigned to imaging equipment by the manufacturer to identify the equipment. The UDI is a system that the FDA continues to implement to identify medical devices through their distribution and use (FDA, 2018).
- *Device category.* Assets are generally organized into categories such as vital signs monitors, laboratory analyzers, and ultrasound machines. This allows further categorization of similar makes/models of equipment.
- *Purchasing information.* This includes purchase order number, purchase price, and purchase contract.
- *Install or go-live date and warranty length.* These help provide information on the age of the equipment as well as manage warranty repairs.
- *Location information.* This includes owning department (or cost center), building, floor, and room location.
- *If the device has an alarm.* This has been added to most CMMS systems to help manage alarm management issues.
- *Risk group (critical/non-critical/other).* This identifies the risk of the equipment as required by regulatory agencies.

- *Operating system, software version, IP address, MAC address, and other network information.* These help manage cybersecurity issues.
- *User and maintenance manuals.* These can be attached to the asset individually or be lined to the specific make/model of the device. Most regulatory agencies require departments to have copies of manuals that define the preventative maintenance (PM) intervals and procedures.
- *Other user-defined fields.* Most systems allow individual departments to identify other fields they would like to add to the system to help identify the asset. These may include data like the RFID tag number assigned to the asset or technician responsible for maintenance of the asset.

Each asset record can then be tied to all of the work (preventative, reactive, or other) done to that asset. The asset history is a rich repository of data frequently used by the department to recommend actions on the device. Types of data analysis that can be done using the device history will be described in the Data-driven decisions and data analytic techniques section later in this chapter.

Work order management

Work orders document all of the work done to assets by the technicians and staff within the department. They can be thought of as a record of all tasks that have to be done in the department. They are generally categorized into PM, reactive maintenance, and other types of work. Regulatory bodies require documentation of all work done, and it is often stated that "if it's not documented, it didn't happen." Work orders are full of data that are useful in data mining and generally indicate the work order type, asset number, requestor of the work, priority, start date, and end date. Work orders can also capture the labor needed to complete the work, parts/materials and other costs to complete the work, test equipment used, and any other notes taken by the technician on work done.

In the next section, we describe the different types of work orders and their purposes.

Preventative maintenance work orders

Once an asset is built, CMMS systems allow them to be added to a plan for PM. This allows a schedule to be set for PM of the system. Systems are generally flexible to the extent that they can plan for several different levels of PM at different intervals. For example, if the system requires an annual maintenance check as well as a battery replacement every other year, two levels of

PM can be created and assigned different frequencies. The schedule can be aligned to a specific month and assigned to a specific technician or group. Systems can also define the length of the maintenance window, which not only provides a start date for the maintenance, but an end date as well.

Once these fields are defined and set, most systems then automatically generate the work orders on the start date for the maintenance and assign them to the technician responsible for the work. These work orders then automatically show up in the work queue of technician and generate a record in the history of the unique asset.

It is important to refer to the user/maintenance manual when setting these PM schedules as all regulatory bodies require maintenance intervals to be adhered to. When changing PM schedules, one must ensure that the length between maintenance activities does not exceed the schedule in the manual, or an interim maintenance must be completed. For example, assume a physiologic monitor is done annually in March. During its life cycle, the device moves to another department, and the technician wants to align the PM with the month that PM occurs in most of the other equipment in the area, which is May. To ensure that the PM is done at least annually, it should be done in March, then May, and then the following May. To ensure that PM dates are known and easily identified, most clinical engineering departments also place a sticker on the device with the last PM date and the next PM date. This indicator not only allows clinical engineering staff to identify when the next PM is due, it also provides data to the clinical staff to aid in identifying assets due for PM.

Most PM work orders will also include a result code. This helps categorize the PM for reporting purposes. Result codes can include:

- *Complete.* This code is chosen if the PM was completed,.
- *Failed.* This code is chosen if the steps during the PM could not be completed because the device did not function appropriately or meet tolerances. Most failed PMs then generate a reactive maintenance work order to repair the device.
- *Could not locate.* If the device could not be found during the PM window, some departments create policies and procedures for staff to follow. This code helps departments track the devices that could not be found and helps identify items for further follow-up after the PM window.
- *In use.* Some devices may in continual use during the PM window. For example, ventilators or infant incubators may be in use continually for weeks or months on end. If the device cannot be removed from

the patient to complete the maintenance, departments create policies and procedures for staff to follow to ensure that patient safety, risk, and regulatory requirements are balanced. Usually departments will tag such devices as "in use," and the full PM procedure is completed once they are no longer in use.

- *Retired/removed.* If the device has been removed from service or retired, this closure code shows that the PM was not completed and that the device is no longer in use. This closure code may also trigger a workflow to remove the asset from service so that no additional PM or reactive maintenance work orders are generated. Assets are never deleted from the database, but generally they are archived should the data be needed again.

Reactive maintenance work orders

Reactive maintenance work orders document all of the repairs done to medical equipment. A reactive maintenance work order can use the same result codes as a PM work order, however more fault codes are used. These codes help categorize why the device failed. This data point allows for additional data analysis that can help identify ongoing or recurring problems with individual assets or categories of assets. Some departments categorize their fault codes as follows:

- *Hardware failure.* This category is used for all failures of any hardware associated with the device including circuit boards, housing, wheels, screens, and so on.
- *Software failure.* This category is used for all software failures such as blue screens, forced restarts, and so on.
- *Accessory failure.* This category is used for all accessory failures such as leads, probes, connection cables, and so on.
- *Battery failure.* This category is used when devices on battery power fail because the battery was not properly charged or did not perform as intended.
- *IT/network failure.* This category is used when the device failure can be tied to an issue with the network, integration, or other IT issue.
- *User error.* This category is used when it has been identified that the device did not function as intended because the user did not properly use the device. Although this may be difficult to determine, sometimes error logs can provide details on which keys were pressed indicating misuse of the device. This type of failure may help indicate which devices require additional training to use.

- *Physical abuse.* This category is used when devices are dropped, smashed, or otherwise abused. This type of failure helps determine which devices, departments, or staff can be additionally trained in the proper use and handling of the equipment. This failure code can also help identify areas for cost savings as the cost to repair this type of failure is preventable.
- *Could not duplicate.* This category is used when the error indicated by the end user could not be duplicated. This helps identify errors where no repair was undertaken, however if a repeated history of this type of failure code is found, it can indicate that further troubleshooting is needed for the device to identify and resolve the ongoing issue.

Other types of work orders

In addition to PM and reactive maintenance work orders, most hospitals also identify other types of work orders to document other work done. The work orders can be categorized into ones that document work done to equipment and other non–equipment-based work. All hospital systems implement equipment-based work, however only hospitals that want to collect all data on labor and productivity collect data on non–equipment-based work.

Equipment-based work order types

- *Initial or incoming inspections.* These inspections are done when a new piece of equipment is delivered and installed in the hospital. They may consist of assembly, calibration, configuration, and function checks. Some regulatory bodies require an initial or incoming inspection to be performed and documented before the first clinical use.
- *Return inspections.* These inspections are done when a piece of equipment returns from an off-site repair, depot repair, or calibration. Similar to the incoming inspection, this inspection ensures that the device is functioning as intended before return of the device to the clinical unit.
- *Recalls/alerts.* These work orders detail the work done by the manufacturer or the technician in the event of a recall, field modification, or safety alert. Documentation should include the notification from the vendor and instructions as to the work done to the device.
- *Upgrades.* This work order documents any software, hardware, or other upgrades done to the device.
- *Incidents/investigations.* When patient safety incidents occur that cause temporary or permanent harm to the patient, equipment that may have

been a factor in the harm is generally sequestered, and an investigation is conducted. This investigation is documented using an incident/investigation work order. All tests conducted on the equipment or reports done by third parties are kept in this record. One must be careful not to put personal protected health information such as patient name or medical record number into these documents. The focus should be on the testing of the equipment.

- *Non–hospital-owned equipment.* Equipment that is not owned by the hospital, such as rental equipment or devices available for trial use, enters the hospital and is used on patients alongside equipment that is owned by the hospital. Inspection of these devices is similar to an incoming or returning inspection, however instead of aligning the work order to the asset number of the owned device, no asset number is defined. Some hospital systems use the device serial numbers or unique numbers used by the rental agency to track these devices. These inspections also ensure that the device in good working order. Patient-owned equipment (such as a phone charger) is, at minimum, visually inspected by the patient care staff to assess proper working order and condition, including signs of hazardous wear. Some facilities may include evaluation by the technical staff.

Non-equipment-based work order types

- *Training.* This type of work order documents the labor associated with any training including on-the-job training, training provided by manufacturers, training provided by third-party servicers, and other organizational training. Appropriately set up, the training work order also allows for reports to be run that help document all of the training individual technicians have received over their career. This report can be helpful to identify training gaps across the department or answer regulatory questions.
- *Meetings.* This type of work order documents all meetings attended. This helps in collecting all data needed to determine the proportion of time spent on technical work and time spent on administrative tasks.
- *Projects.* Most CMMS systems allow departments to create projects to collect all work orders associated with the project into a grouping. This helps identify which projects the department is working on and how much time is spent on a particular project.
- *Other.* Specific work order types may be created for a department's specific need. CMMS systems are generally flexible enough to allow

creation of specific other work order types. Departments should think carefully about the creation of too many work order types, however, because this may affect data categorization and reporting.

Labor tracking

Each work order within the CMMS system also has the ability to track the amount of time the team member works on that work order. Time can be added in several ways including regular time, overtime, and travel time. Analysis of this type of time tracking can be helpful in several ways.

Time tracking in general can provide data to help analyze the staffing needs of the department. In general, 30% of time spent across the department is spent for repair activities, 30% to 40% is for PM, and the other 30% to 40% is for other work order types. The number of hours of PM, repair, and other work order types gives insight as to how many staff will be needed to support the equipment. Additionally, analysis of the percentage of work spent in different categories helps identify possible changes needed in the equipment maintenance management program. For more analysis insight, see the Data-driven decisions and data analytic techniques section later in this chapter.

Overtime tracking can identify the types of repairs/maintenance done using higher-cost overtime hours. Review of these types of work orders informs leadership on changes that might be made to balance the needs of the organization with cost-containment strategies. Analysis might also inform the leadership of staffing changes needed to cover a large number of repairs or maintenance done after hours.

Travel time tracking can help inform leadership as to how much time technicians spend in a vehicle traveling from one site to another. This helps identify opportunities for grouping work together to minimize this non-repair time. It can also help justify resources (both staff and vehicles) needed for the servicing of all sites.

Finally, analysis in time tracking can also help improve staff productivity and quality. Setting and monitoring goals for how many labor hours that need to be documented per an 8-hour shift can help drive appropriate documentation and productivity by staff. Additionally, standardization or normalization of work order labor times for makes/models of equipment can help reduce variation in labor. For example, a PM procedure on a specific make/model of dialysis machines may be 1.2 to 0.3 hours. Staff that are taking less time than this may not be performing the entire procedure correctly, which may affect quality. Staff that are taking more time than this may not be adequately trained to complete the PM. Generating these norms requires appropriate data entry in each work order consistently across the department.

Parts management

Each work order type also allows for allocation of specific parts by number and cost. To allocate these parts to the work order, most CMMS systems require a record to be built for each part number. This record includes the manufacturer, distributor, part number, cost, and which equipment the part is used for.

Most departments track large and medium-sized/cost parts because tracking and analysis of these parts have an impact on the budgeting and measurement of the total cost of ownership of an individual device. However, tracking the small items (e.g., screws, bolts, nuts, washers) becomes too onerous. The specific type and category of parts managed by the department will differ from department to department, but the decision should be thought out proactively before implementation as it will affect the accuracy of cost for repair and PM as well as the budget.

Once parts records are built, the individual part number can be added to any work order. This allows a record of all parts used during PM or repair of the device. Costs for those parts are also allocated automatically, which helps identify the cost of the PM or repairs over the life of the equipment.

Some more advanced CMMS systems also allow integration of the parts module to the hospital's procurement system. This integration can allow for parts to be ordered seamlessly if they are not in stock in the department. Parts ordering can be set up as proactive or reactive. Proactive parts ordering occurs when minimum parts numbers or pars are set in the system. When the parts are allocated on work orders and decremented from the inventory, the system identifies that the minimum number for a specific part is not met in inventory and automatically generates a request for reorder. This request can then be transmitted to the procurement system and ordered. This proactive system must ensure that all parts in inventory are routinely checked to match the quantity in the system so there is no mismatch between what the system believes is in stock and what actually is in stock.

Reactive parts ordering is used for parts that management does not want to keep in inventory because they are not used often enough or are too expensive to keep. These parts are only ordered when needed for PM or repair.

The integration between the CMMS and the procurement system can also be configured to provide the staff ordering information on the part, such as the purchase order, delivery date, and invoice, which can help them manage their work.

Contracts management

Most CMMS systems also have a contract management module. Most departments are not able to repair all equipment internally, so sometimes departments contract with the OEM or a third party to perform some or all services on a device. The start date, end date, terms, and coverage can be managed in the contract management module. For departments with many contracts, it is often difficult to keep start dates and end dates organized so that coverage does not lapse and to be informed as to what is covered under the contract and what items need to be purchased off-contract. A CMMS contract record includes basic information such as the contract number from the vendor, a copy of the contract, and the start and end dates. The dates can be used to trigger notifications 30, 60, or 90 days before the end date so leadership can start the process of contract renewals. Additionally, information on the type of coverage can be detailed including items such as a parts-only contract, repair-only contract, PM-only contract, or full contract with parts, repair, and PM included. Details on labor rates or parts discounts for non—contract-covered services can also be added. This helps communicate to the team which items should be included as part of the contract and which items need to be paid for separately. Finally, the items covered can be linked to the asset records so that when technicians pull up the asset record, they can clearly see that the device is under contract.

Reporting

When implementing a CMMS system, one should start with defining the reports required and work back to identifying the data required for these reports. With the data identified, one can then create the fields and work order types required to populate the reports. It is impossible to create a report if the required data are not available.

CMMS systems can provide all kinds of pre-created reports that use the basic data that are collected in work orders. Most systems not only offer these standard reports, but also offer the capability for systems to build or create their own custom reports. Basic reports generally fall into a few categories:

- *Work order reports* detail the number and types of work orders, result codes, and closure codes.
- *Asset reports* include detailed asset histories, or *device category reports* bundle all assets in a specific device category. These reports can be leveraged to determine how much time or money is spent on an individual asset or set of assets. There may also be reports that track asset warranties and life cycles as well as provide inventories by department or location.
- *Schedule reports* detail the PM schedules of devices. These can generally be run by month maintenance due, or be used to determine what is overdue.
- *Materials or inventory reports* describe which items are currently in inventory, determine parts utilizations, and identify items that need to be ordered. There may be reports to detail idle inventory or other inventory that is not frequently used that may be identified for salvage or return to the manufacturer.
- *Account reports* are built for the purpose of providing accounting information generally for billing purposes. These reports detail work done organized by department or account and the associated costs. They are used both to inform departments of work done in their area and to create a billing invoice if the department charges for the work.
- *Worker reports* include lists of open work orders, work order completion percentages, workload measurements, and worker productivity (hours documented/hours worked).
- *Contract reports* include lists of all current contracts, reports of contracts expiring or expired, and measurements or analysis of cost of maintenance for devices under contracts.

Most CMMS systems also have the ability to allow for creation of custom reports. Because these databases are relational, all of the data elements are available in several tables. Using the tables and elements, any report pulling any of the data available can be created.

Medical device interoperability

At the time of publication, about one-third of medical devices at a typical hospital are able to connect to the hospital network, although communication protocols may vary. Utilizing this connection, interoperability is generally the communication between highly specialized devices, resulting in the transfer of patient data to the electronic medical record. The medical devices, when connected, continue to function as intended while sharing information. The fundamental effort in interoperability is data capture: data are collected by a medical device, captured for transmission, stored, and then available for retrieval through the electronic medical record. Before this interoperability, healthcare providers were required to evaluate the information provided by the device at the bedside and manually enter the values into the medical record.

Automated data capture from medical devices can be challenging. Issues with this process include the following:

- *Devices can collect enormous amounts of data.* For example, physiologic monitors gather information about the electrical activity of the heart with sampling rates that far exceed useful information. Thus, decisions must be made that determine appropriate sampling rates to archive meaningful information. Engagement with medical practitioners to seek consensus on data storage is essential. Efforts to avoid "drowning" sentinel events with mountains of mundane information are critical to appropriate data storage.
- *Computer systems cannot easily identify erroneous information.* In traditional data-gathering processes using a human interface, a technician would repeat a measurement if something interferes with an accurate reading. For example, if a patient moves or removes a sensor, the meaningless data would not be archived because the clinician would not record it. However, when automated systems archive data, invalid readings may be stored. Automated data-retrieval tools must be "taught" to discern high-value information.
- *Standard data collection, storage, and transmission techniques vary widely between devices.* Each device has its own method to transmit data (USB, RS232, Ethernet, wireless transmission).
- *Data formats.* The number of unique formats and/or protocols used to transmit data between devices and across systems is estimated to be more than 100.

To facilitate the connection of devices to the hospital network given the wide variety of protocols and communication methods, middleware tools and software, called *medical device data systems (MDDSs)*, transfer medical device data, often by converting formats. MDDS tools are another component of the highly complex medical device system.

In 2019, AAMI published the Technical Information Report, TIR75: *Factors to consider when multi-vendor devices interact via an electronic interface: Practical applications and examples*. This guidance document provides information associated with the safety risk evaluation of interoperative functionality as well as the important considerations needed when these connections are made.

Project management principles, as applied to clinical engineering

Projects with specific goals are integral to clinical engineering. The implementation of a new imaging modality or the acquisition of new anesthesia machines are examples of common projects that demand excellent project management. The overarching goal for clinical engineers is to manage a project so that the goal is met on time, on budget, and within specified performance parameters. Information management—communication—is key to successful project management. Clinical engineers must determine the amount of information needed by each stakeholder and deliver it when needed. Time management through scheduled milestones is vital and is often facilitated by software. Project progress tracking with schedule adjustments is implemented when challenges need to be managed through effective communication and leadership.

Organizational skills are key to successful project management. Clinical engineers must document activities thoroughly and be able to locate information and materials consistently. Meetings must be effectively coordinated and hosted, with strong leadership that appropriately manages team members. The healthcare environment can be riddled with power distance challenges—that is, inequities in status among individuals are integrated within patient care. Hierarchical relationships, such as between doctors and nurses, can sabotage project success if not managed carefully. Stakeholders must be valued and contributions incorporated into project implementation regardless of stature or rank. Effective project managers value the input of everyone involved and affected by project outcomes. Clinical engineers are most successful when they manage and organize the information received from stakeholders, driving the project to successful completion.

Clinical engineers must effectively communicate throughout the project in many ways, including visual representations, brief status updates, executive summaries, and detailed reports. Clear and concise language and effective word choices can promote deeper understanding and stakeholder support.

The principles of effective technical communication, including email usage and training material creation, are critical to successful project management.

Ongoing support planning including service agreements

Background

Support planning occurs at a few separate levels. Most departments have high-level support plans from the department level view that take the form of service-level agreements (SLAs) and operational-level agreements (OLAs). These high-level plans may define broadly how the clinical engineering department supports the hospital or clinics, but they do not provide details on support of individual types or categories of equipment or individual pieces of equipment. These device- or category-specific support plans provide much more detail on how the specific device is supported, including whether specific portions of the service are completed by the in-house staff or outsourced to a third party.

Service-level agreements

An SLA is a document that defines services provided between the clinical engineering department and the hospital organization. Particular aspects of the service provided in this document include the following:

- *Department structure* defines the number and type of staff and management, staff location, and contact information.
- *Responsibilities* are areas for which the department is responsible and areas for which the department is not responsible.
- *Availability* includes department hours for on-site support versus on-call services; how to obtain services including normal processes for how to receive services; and processes for escalations.
- *Responsiveness* specifies the expected turnaround times for on-site and on-call services.
- *Quality* includes how the department will maintain quality support for the organization (Kearney & Torelli, 2011).

Some clinical engineering departments have different SLA agreements for different types of hospitals (e.g., academic medical centers, tertiary care hospitals, community hospitals, critical access hospitals) and different types of clinics (e.g., procedural clinics, primary care clinics, urgent care settings) because the support hours, availability, responsiveness, and user needs are different for these various locations. Tailoring the services the clinical engineering department provides to various entities allows the flexibility to provide only the services needed to different sites, which allows for optimization of resources across the health system.

Operational-level agreements

An OLA defines the interdependent relationships, roles, and responsibilities across various departments within the organization in support of the institution. These documents are helpful to define every group's accountability in the process of support. The accountabilities include the process for handoffs among teams and the time frame for delivery of their services. The objective of the OLA is to present a clear, concise, and measurable description of the internal support relationships and describe how support across groups occurs. Although most OLAs focus only on support groups, the end user should be included as a group as end users have a specific role and responsibilities to perform in supporting their own devices.

Device-specific support plans

Although departments create high-level SLAs and OLAs, these documents are very broad and generally relate to the entire department and not to specific device categories or types. Often, specific plans for individual device types and categories are needed because of specific needs of the department, the users, or the technology.

Creating a support plan for a specific service or device generally follows three steps:

1. Determine the scope.
2. Determine the roles and responsibilities.
3. Document the support plan.

Scope

During the scope development phase, a team is built that includes individuals from all areas of support and all end-user departments that have requirements for support. The team then discusses all aspects of the need. This includes the *who, what, where, when, why*, and *how* of the system needs. For example, if a hospital is installing an imaging system, the team would include the imaging staff/leadership, clinical engineering staff/leadership, information technology teams including PACS teams, medical records teams, cybersecurity teams, and perhaps even the vendor.

A discussion would then include who owns and operates the system, what the imaging system does, what types of patients are being seen, what hours the imaging device will be used to service patients, where the device is going, when it will be set up, how it will be used, how it will connect to the various IT systems including PACS and medical records, how the data will flow, and who will access them. Discussions around ongoing support of the system would also be conducted, including which type of support is needed (e.g., phone, on-site, on-call), when/who will conduct maintenance, expected turnaround times during working and nonworking hours, when patching can be done, what types of upgrades are expected, what types of monitoring are done, and so on. End users and support groups would also discuss their workflow processes, regulatory requirements, and any other business needs.

This step allows cross-disciplinary teams to understand each team's processes and people and be on the same page with expectations and business needs. Once the scope is fully defined, the team can then discuss and define the roles and responsibilities for each of the team members for all support processes needed to meet business needs.

Defining roles and responsibilities

The scope definition portion helps identify the people, process, and technology needed to support the system both for the installation and for ongoing support. This step takes these needs and matches them with existing roles and responsibilities within and external to the organization. These roles and responsibilities are then memorialized in a documented support plan that is signed off by all parties.

All of the processes and tasks identified in the scope definition should be broken down into steps and the people responsible and accountable for the step identified. For example, if the staff expect to call the standard helpdesk for downtime issues, the helpdesk personnel are identified as the people taking these calls. The hours of the helpdesk should match the expectations for the end user so that support expectations match the reality. Additionally, this identifies the need for scripts and training to be developed for the helpdesk staff to deal with the most common type of issues predicted to occur. Other examples might identify the system administrator who would be responsible for running reports, adding users to the databases, and training frontline staff on using the system.

When defining the roles and responsibilities, care also should be taken to identify support needs that the vendors (either OEM or third-party) also need to provide. When deciding whether to insource or outsource these support needs, the team needs to balance the cost against the identified scope, roles, and responsibilities. To help determine if the support should be outsourced, the team should ask themselves several questions including the following:

- Are the needs of the end users more than can be provided with existing support models?
- Are existing staff able to be trained for support? Do they have enough bandwidth to support this new system?
- Is this standard equipment or on-off/specialized equipment?
- How much will it cost to support over the lifetime of the system if the support is insourced versus outsourced?

Once the decision is made to insource or outsource all or portions of support of the system, a single owner should take responsibility for all vendor management functions for the support plan.

Draft support plan

Once all processes are mapped and people are identified in their roles and responsibilities, all of these decisions and workflows must be documented in a support plan. The following topics should be included in this plan:

- *Document history and revisions* cover the history of the support document and how many times/when it was revised.
- *System application overview/vendor contacts* discuss the system/processes at a high level including the vendor and how to contact the vendor.
- *Technical information* covers all of the technical information, including but not limited to:
  - server/appliance overview;
  - database overview;
  - interface overview;
  - VPN/firewall overview;
  - network overview;
  - workstation overview;
  - application overview;
  - storage overview;
  - biomedical equipment overview; and
  - technical/application/architecture diagrams.
- *Data flow overview* discusses how the data flow through the system and identifies the values/names of variables that are transferred.

- *Roles and responsibilities* identify all parties by role and what they are responsible/accountable for, including:
  - IT service areas;
  - clinical engineering;
  - end user; and
  - vendor.
- *Support processes* usually utilize a flowchart to describe the top 5 to 10 items/processes/faults that may need support and how support is planned to occur.
- *Knowledge articles* provide general "how-to" information shared with staff on what to do in certain situations. This information is generally used to train new staff in support of the system.
- *Sign-offs* provide a page for signatures of all parties that agree to the support plan.

The documented support plan should then be kept in a place that can be accessed by all parties responsible/accountable for support. Plans to train staff on support procedures should be created and completed before the system goes live. Ideally, initial conversations around support and decisions on insourcing versus outsourcing should be decided before purchase so that support needs can be negotiated as part of the overall system purchase. If timely decisions are not made before purchase, vendor negotiations can be done separately for support, but this reduces the leverage to negotiate lower costs or support processes needed by the end user.

Vendor contracting and vendor management

Working with vendors is an integral part of the day-to-day responsibilities of clinical engineers. Usually there are two different ways interactions occur, before purchase of the device/system and after purchase. During pre-purchase activities, the interactions with vendors are usually around vendor selection and contracting. After purchase, the interactions include implementation and long-term support.

Pre-purchase vendor contracting

The vendor selection processes, including Requests for Information (RFI) and Requests for Proposals (RFPs), were discussed in Chapter 2, Healthcare technology basics. After these processes, the device/system make/model is generally selected by the internal stakeholders. After the vendor selection is made, the clinical engineering staff, IT staff, and end users then work with the supply chain team to contract with the vendor. The vendor negotiations can include the purchase price for the system as well the implementation scope of work and support costs, if the decision was made to use the vendor for the implementation and support of the device/system. However, cost should not be the only item discussed during the contracting phase. If there are disposables for system/equipment, these costs should also be negotiated at the time of purchase. This is also the time to include any terms and conditions the hospital/health system also require the vendor to accommodate. These may be items including but not limited to the following:

1. *Warranty information* including items covered, length of the warranty, hours the warranty service will be provided, whether PM will be done by the vendor during the warranty period, and any exclusions to the warranty. There should also be a set way to determine the start of the warranty (delivery of the system versus first patient use of the system).
2. *Inclusion of users and maintenance manuals* and the form in which the manuals are provided (electronic versus paper).
3. *Third-party designations*, which are statements indicating if it is required for the vendor to get approval from the hospital before using a third party to install, service, or maintain the equipment in question.
4. *Vendor training expectations* for hospital user and support staff.
5. *Vendor installation and repair expectations*, including how the vendor is expected to check in to the hospital/health system, requirements for vendor support staff for training/health screenings, if new or refurbished parts should be used, how reports/bills will be delivered to the hospital/health system (electronically/on paper, in a certain number of days, or in a specific format), and if prior approval must be obtained before performing the repair or change.
6. *How the vendor will notify the hospital* and handle recalls/alerts/field modifications and changes in operating, reprocessing, cleaning, and maintenance procedures.
7. *Obsolescence protection* including how the vendor will notify the hospital of equipment reaching end of life or becoming obsolete, how the vendor will reveal planned upgrades or introductions of new-generation systems, and how long the vendor will continue to support existing systems after new systems are placed on the market.
8. *Early termination clauses and procedures* including notice and payment.
9. *How additions/deletions from the contract will be handled* including start dates, end dates, and payment terms.

10. *If the hospital wants a semiannual/annual business review* with the vendor and what the review will include.
11. *A system performance guarantee* including uptime guarantees and costs or penalties for not meeting these guarantees.

The supply chain/legal contracts team should also evaluate contract negotiations to review terms including indemnity and out clauses and include standard business associate agreement terms and conditions. IT should also review terms to ensure that any cybersecurity or other required IT terms are included.

This inclusion of terms and annual quality assurance review to ensure that the vendor meets these terms satisfies one of the Joint Commission requirements related to vendor quality.

Post-purchase vendor management

After the initial contract is signed, the management of the vendor begins. When defining the accountable/responsible parties during the support planning, some tasks/processes may be the responsibility of the vendor or other third-party entity outside of the organization. When vendors are identified as part of the support processes, there also needs to be an owner identified to manage the vendor. This vendor owner is responsible for managing the vendor during the implementation and ongoing support of the system/equipment. This vendor owner is the main point of contact for the vendor and often is a liaison between the end user and the vendor.

During the implementation phase, vendors may provide a project manager and implementation scope of work to help the hospital/health system get the system or equipment up and running. The vendor owner is responsible for working with this vendor project manager to get the system/equipment up and running. This may include traditional project management tasks or even acquiring an internal project manager, depending on the complexity of the installation.

Post-installation, the vendor manager is responsible for overseeing ongoing contract and support management from the vendor. This includes contract renewals, contract amendments, annual vendor quality reviews, payment issues, and any other vendor management issues that need to be resolved.

Holding the vendor accountable to expectations requires a proactive discussion about what the expectations are and a measurement tool to assess the vendor's performance. The terms and conditions placed in the contract should include the vendor expectations, as this is a contractual way to hold vendors accountable. A tool such as a vendor scorecard can then be developed to measure vendor performance (Jacques & Watson, 2017). Vendor scorecards can be developed for implementation tasks or ongoing support tasks.

Fig. 3.2 shows an example of a simple scorecard for a vendor who was contracted to provide education documentation and training to three groups of staff. As you can see, each requirement, measurement, and result

Requirement	Measurement	Result/outcome	Next steps
Vendor follows policies on badging and vendor credentialing	• Audit of vendor credentialing software • Assessment of badges when vendors are on site	All on site vendors were credentialed in vendor credentialing software as measured on August 1. However on one occasion (June5) hospital observed John Smith entering unit without his badge. Staff discussed with him and issue was resolved immediately	Please review vendor credentialing policy with vendor's staff and assure all staff is credentialed. Follow-up review will be conducted next quarter
Vendor required to meet June 1 deadline for creation of staff education	• Review of schedule and deliverable • Feedback from	Staff education presented to team on June 1. Staff review of documentation show it to be complete and accurate.	None
Vendor to train all staff by July 1	• Audit of staff training, schedule and quality	Education sessions scheduled on June 3, 19, and 22. Vendor was on site and on time for sessions on June 3, but was late for the session on June 19 and did not show up for June 22 session. Training not completed as of August 1	Vendor to reschedule training for teams that were to be trained on june 19 and 22. Training will be completed by August 31. Failure to complete this timely will be considered a breach in contract
Vendor provided invoices 30 days after training	• Receipt of invoice	Invoice not paid as work has not been completed	

Figure 3.2 Sample vendor scorecard for a scope of work for a vendor to create education and train staff (Jacques & Watson, 2017).

is detailed. When completing the scorecard, if follow-up or next steps are required, they should be documented as well.

Creating these scorecards and sharing them with vendors to set expectations help proactively manage the vendor and ensure successful support of systems. Regular ongoing assessments of the vendors also develops a relationship between the hospital and the vendor.

Data-driven decisions and data analytic techniques

The availability of large amounts of data and the use of this information has transformed some activities associated with healthcare technology management. For example, one hospital may have a few dozen of one type of monitor. The repair history associated with this model in the experience of the one hospital may offer a small data set of issues and solutions. However, when a large number of facilities share their repair histories, a larger and more statistically significant data set can drive a deeper understanding of common issues associated with the particular device. Pooling repair information is only one example of the use of large data sets to drive actions such as repair parts storage and PM activities. The analysis of data sets generally drives the process of evidence-based decisions. This contrasts with anecdotal evidence or a "gut feeling."

Biomedical informatics, also known as *healthcare information technology*, *healthcare informatics*, or *biohealth informatics*, is an emerging field that focuses on the management and organization of biomedical/patient data that are often aggregated and analyzed to improve patient care. Large data sets associated with patient care, including information associated with the electronic medical record, may be referred to as "big data." The analysis of these huge sets of information requires hardware and software tools and statistical analytics.

One application of data analytics in healthcare is clinical decision support (CDS). This tool guides the medical professional in decision making based on the analysis of large sets of patient information. Diagnosis (such as sepsis alerts), treatment decisions, and interventions (such as medication selection) can be driven by the analysis of large data sets to offer historical references. As the electronic medical record usage expands and more physiologic data are gathered and stored, the data sets may be diverse enough to offer highly reliable assistance to clinicians. The future of evidence-based decision making is potentially substantial although currently limited.

Financial management of medical technology

Clinical engineering departments are responsible not only for the repair and maintenance of equipment but managing the costs related to these repairs and maintenance. Proactive departments also use the data in their CMMS systems to aid in planning for capital replacements of medical devices. Managing the finances of the department requires day-to-day management of costs as well as planning future capital and operating expenses. Finance departments within the hospital/health system use Generally Accepted Accounting Principles to define how financial management for the hospital/health system occurs. In general, operating expenses can be separated into two categories: revenue and expenses. Revenue refers to money collected for services rendered, while expenses refer to money spent to render these services. Revenue and expenses are organized into categories that are budgeted for and measured against. Capital expenses are usually defined as assets (physical or virtual) that have a useful life of greater than 2 years and cost greater than $5000. Most hospitals/health systems use an annual process to plan for both capital and operating expenses. Once budgets are approved, the actual costs are measured against the budgeted amounts.

Budget management

Clinical engineering operational budgets are generally very expense based; very few departments generate revenue. Those departments that have revenue either sell their services to third parties for repair/maintenance of sites outside their hospital/health system or use the revenue line for capturing the payments made for retired equipment that is sold to a third party.

Clinical engineering department expenses can be categorized into a few large categories including the following:

- *Salaries/personnel costs* include salaries of staff, costs for overtime, on-call pay, paid time off, budgeted raises, and employee recognition/rewards expenses.
- *Supplies* include cleaning supplies, office supplies, calibration gases, minor equipment such as tools or test equipment, and other maintenance supplies.
- *Repair costs* include all costs for parts purchased to repair devices and any kits or parts needed for PM.
- *Outside-purchased services* include costs for departments to use a third party to repair or preventatively maintain equipment, calibrate test

equipment, or contract with to provide any services to the department including maintenance services agreements.

- *Training/travel costs* include costs for registration, books, educational sessions, conferences, and training for staff. If the department supports sites they need to travel to, this may also include costs to travel to and from all sites that are supported.
- *Other costs* that may be budgeted are computer software costs for CMMS systems, uniforms, communication (phones/pagers) costs, equipment rental or lease costs, shipping costs, and depreciation costs for any capital equipment owned by the department.

During the budgeting process, most hospital systems will use historical costs over the past few years to predict the future year's costs. Adjustments are then done to the predictions related to new expenses or cost savings the department expects to occur in the next year. Once budgets are finalized, they are generally approved by leadership before the beginning of the fiscal year. During the fiscal year, the actual expenses are compared to the budgeted expenses to monitor how the department is performing. Leadership can then use this comparison to help decide on how to move forward with repairs and expenses on a day-to-day basis.

Capital planning

In additional to annual operational budget planning, clinical engineering may request any capital equipment needed to support the department. This may include new test equipment that is needed to support acquisitions by the hospital/health system. This request process is usually done annually and in conjunction with the operating budget.

Clinical engineering can also play a role in helping other departments plan their capital requests. Using data from the CMMS system, clinical engineering can provide data on the life of existing equipment, cost to maintain that equipment, and recommendations on whether to replace existing equipment. Clinical engineering can also help identify replacement equipment including vendor recommendations, support with RFI/RFP processes, and discussions about the long-term costs of new equipment.

Most hospitals/health systems require justification of new capital purchases including measurements of total cost of ownership or return-on-investment (ROI) analysis.

Return-on-investment analysis

ROI analysis helps hospitals/health systems understand how effective an investment is. The ROI is calculated by taking the benefit or return of the investment and dividing it by the cost of the investment. The result is shown as a percentage. For example, if the hospital wants to purchase a new ultrasound scanner for $100,000 and expects it to generate a net new patient revenue of $120,000 each year, the ROI in year 1 is 1.2%. This would be a good investment as the cost of the equipment will be paid off by the revenue in less than 1 year.

When determining what capital the hospital/health system will fund in the future year, an ROI analysis helps prioritize capital that increases revenue while managing costs and limited resources.

Clinician connections including communication and collaboration

Clinical engineers have a very direct connection to clinicians (physicians and nurses) as the medical equipment is the direct interface between the clinician and the patient. There are several typical projects led by clinicians that clinical engineers become involved with including alarm management, standardizing devices/disposables, and other research projects. Some topics are of broad concern, appearing in journals and trade publications.

Alarm management and secondary device notification

Hospitals rely on alarm-equipped medical devices to provide appropriate care to patients. Alarms are designed to alert caregivers to events that need intervention. However, with all the various equipment in the patient care area, the quantity of alarms has become enormous. Alarms have now become pervasive and have become a hazard instead of providing value to clinicians because of their quantity. Under the Joint Commission's National Patient Safety Goal on Alarm Safety (NPSG) (The Joint Commission, 2013), hospitals are tasked with implementing alarm management protocols and educating clinical staff on alarms. There are many published recommendations on how to reduce non-actionable alarms, including the *Clinical Alarm Management Compendium* (AAMI, 2015), but all of the recommendations start with the creation of a cross-functional team including clinical staff and clinical engineering. The team then works together to identify and implement ways to reduce the clinical alarm load.

This may include changing default parameters on devices and educating staff on setting alarm limits based on patient populations or individual patient baselines (AAMI, 2015; Jacques, 2016; Cosper et al., 2017).

There is also technology available called *middleware* that will take alarms from various devices, filter them, and send the filtered alarms to a secondary device (usually a smartphone or pager) (Jacques, 2017). This allows alarms to be sent to the caregiver directly responsible for the patient's care. The design, implementation, and long-term support of this middleware is also done by a cross-disciplinary team including clinicians, IT, and clinical engineering.

Device/disposable standardization

Clinical engineering staff also work collaboratively with clinicians and supply chain staff when setting or standardizing devices or disposables. These standardization efforts are usually ways to optimize and streamline processes and reduce costs in a health system. For example, standardizing ECG cables allows the supply chain to leverage vendors for a volume discount, while allowing nurses to use the same cable in any unit they work in, which minimizes training. When determining which cable to choose, clinical engineering staff can work with the clinical staff to evaluate available cables and select the one that is most reliable and functional while also being cost-effective.

All manner of equipment from thermometers to magnetic resonance imaging machines can be standardized, which also aids in standardization of the disposables that are needed for these devices. This standardization reduces the overall cost for supplies and allows leverage for negotiation of reduced costs as a result of volume discounts. The number of repair parts is also reduced as the number of different makes/models supported is reduced. Finally, training on how to use the equipment both by clinical staff as well as support staff is optimized as staff only have to learn one type of device.

Full standardization is generally not accomplished because of specific needs to certain service lines, but the goal to standardize as much as possible helps optimize clinical and support processes.

Research

Clinicians and clinical engineers can also collaborate on all manner of research activities including translational research projects, which take inventions, devices, or processes and transition them from basic research to clinical practice. This research activity generally occurs in academic medical centers and other hospital/health systems that have research as part of their mission. Most community hospitals or critical access hospitals have little or no ongoing research activity.

Emerging technologies and technology transfer into healthcare settings

One of the most entertaining activities for a clinical engineer is to read books published in the 1980s and note the predictions for healthcare technology of the future. Rarely are the expectations on target, and some are incredibly wrong. As a result, it can be frustrating to see too far into the future of healthcare technology. That said, a few innovative trends likely to enter the clinical setting are worth noting.

- Interconnectedness of devices to the electronic medical record will drive innovation and change as technology will better record patient data without humans writing numbers onto a slip of paper and typing them into a computer database. Once interconnected, devices may offer some opportunities for closed-loop control to improve therapeutic actions.
- The use of three-dimensional printing utilizing many different materials is likely to affect a wide variety of patient care interventions, including customized components for treatments.
- Nanotechnology and nanomaterials offer treatments and interventions on a new and microscopic scale. In addition to nanoparticles that are specifically designed to diagnose and treat disease, tiny-scale electronics can bring monitoring and evaluation into the body in new and exciting ways.
- Artificial intelligence (AI) may drive medical care as well as device design in the future. Medical professionals are exploring CDS as an application of artificial intelligence to identify possible changes in patient status before clinicians can detect them. In addition to patient care, the analysis of historical data as well as predictive algorithms may shape device maintenance and support decisions. AI may offer improved clinical engineering efficiency but will depend on the optimal use of accurate service data. AAMI, in collaboration with several other groups, published a white paper that addresses AI and machine learning. This paper is available for free download (AAMI, 2019).

Engineers and scientists work in research laboratories in academic institutions, corporate environments, and government facilities to identify innovative ways of improving medical care with technology. Creativity has offered unpredicted imaging tools, therapeutic devices, and other technologies that cannot be predicted but will be implemented and managed by clinical engineers of the future.

Abbreviations

CDS clinical decision support
CMMS computerized maintenance management system
INCOSE International Council on Systems Engineering
MDDS medical device data system
NPSG National Patient Safety Goal on Alarm Safety
OEM original equipment manufacturer
OLA operational-level agreement
PM preventative maintenance
RFI Request for Information
RFP Request for Proposal
ROI return-on-investment
SLA service-level agreement
UDI unique device identifier

References

AAMI. (2019). *The emergence of artificial intelligence and machine learning algorithms in healthcare: Recommendations to support governance and regulation.* Retrieved from <www.aami.org/ai_paper>.

Association for the Advancement of Medical Instrumentation. Clinical Alarm Management Compendium (2015). Retrieved from <http://s3.amazonaws.com/rdcms-aami/files/production/public/FileDownloads/Foundation/Reports/Alarm_Compendium_2015.pdf>.

Cato, W. W., & Mobley, R. K. (2002). *Computer-managed maintenance systems: A step-by-step guide to effective management of maintenance, labor, and inventory* (2nd ed.). Woburn, MA: Butterworth-Heinemann.

Cosper, P., Zellinger, M., Enebo, A., Jacques, S., Razzano, L., & Flack, M. N. (2017). Improving clinical alarm management: Guidance and strategies. *Biomedical Instrumentation and Technology, 51*(2), 109–115.

Jacques, S. (2016). Breaking barriers to patient-centric alarm management. *Biomedical Instrumentation and Technology, 50*(3), 181–183.

Jacques, S. (2017). Factors that affect design of secondary alarm notification systems. *Biomedical Instrumentation and Technology, 51*(s2), 16–20.

Jacques, S., & Watson, A. (2017). Proactive vendor management for healthcare technology. *Biomedical Instumentation and Technology, 51*(2), 116–119.

The Joint Commission. (2013). *The Joint Commission announces 2014 National Patient Safety Goal.* Retrieved from <http://www.jointcommission.org/assets/1/18/JCP0713_Announce_New_NSPG.pdf>.

Kearney, K., & Torelli. (2011). The SLA model. In P. Wieder, J. Butler, W. Theilmann, & R. Yahyapour (Eds.), *Service level agreements for cloud computing* (pp. 43–67). New York: Springer.

National Academy of Engineering (US) and Institute of Medicine (US) Committee on Engineering and the Health Care System; Reid, P. P, Compton, W. D., Grossman, J. H., & Fanjiang, G. (Eds.). (2005). Building a better delivery system: A new engineering/health care partnership. Washington, DC: National Academies Press.

President's Council of Advisors on Science and Technology. (2014). *Better health care and lower costs: Accelerating improvements through systems engineering*. Washington, DC: Joint Commission Resources.

US Food and Drug Administration. (2016). *Applying human factors and usability engineering to medical devices—February 9, 2016*. Center for Devices and Radiological Health.

US Food and Drug Administration. (2018). *Unique device identification system (UDI system)*. Retrieved from https://www.fda.gov/medicaldevices/deviceregulationandguidance/uniquedeviceidentification/>.

CHAPTER 4

Safety and systems safety

Introduction

The image illustrating the scope of the healthcare technology management (HTM) profession, presented in Chapter 1, The profession, highlights patient safety at the highest level of the diagram. Patient safety is a shared goal of many members of the healthcare team. For example, the Hippocratic Oath utilized by physicians emphasizes "first, do no harm." Thus, clinical engineers are well-positioned to advocate for patient and staff safety and are interwoven in all tasks. This chapter focuses on the myriad of guidelines, laws, and best practices that help support patient and staff safety in a variety of environments and their associated technology.

Regulations, regulatory bodies, codes, and standards

Healthcare delivery is regulated by a variety of agencies and organizations. This complex landscape is driven to incorporate best practices through both law and guidelines, with the strength of compliance varying based on the oversight of the group, code, or regulation. A *regulation* is a best practice, identified by expert groups, that is generally required by law. For example, the Health Insurance Portability and Accountability Act (HIPAA) is a law that contains many regulations. Failure to comply with regulations can result in legal concerns. Regulations are generally created through a collaborative process involving organizations (e.g., the American Hospital Association) and a public comment period. Organizations often lobby to shape regulations, guiding their creation in a way that is viewed as aligned with the purpose and goals of a group.

The US Food and Drug Administration (FDA) is involved in regulations associated with medical devices. The FDA oversees the approval of devices before they are available for purchase and after the devices are involved in patient care. The Center for Devices and Radiological Health (CDRH) is the division of the FDA associated with medical devices. The Occupational Safety and Health Administration (OSHA), a division of the US Department of Labor, is associated with regulations that protect

Introduction to Clinical Engineering
DOI: https://doi.org/10.1016/B978-0-12-818103-4.00004-1

workers. The Clinical Laboratory Improvement Amendments of 1988 (CLIA) are associated with regulations for clinical laboratories in hospitals. The Nuclear Regulatory Commission (NRC) is involved in regulating the nuclear medicine specialties including gamma cameras.

Within the United States, each state has a board of health that provides a state license to deliver services. Fundamentally, the institution is governed by the State Board of Health and must comply with state regulations to remain open. State surveyors typically review hospitals once per year and may revoke a license based on findings.

Codes are detailed guidance recommendations that may be adopted as law. Some examples of groups that create codes (through a robust process involving many steps and stakeholders) are the National Fire Protection Association (NFPA), which is focused on electrical and fire safety, and the International Code Council (ICC), which crafts building codes. Other examples include:

- *Centers for Medicare and Medicaid Services (CMS).* Health insurance coverage for older Americans and those who cannot afford health insurance is managed by the CMS, an agency of the US Department of Health and Human Services (HHS). To receive payments for patients covered by Medicare and Medicaid, hospitals must comply with the recommendations set forth by the CMS. The CMS *interprets* codes and regulations, offering guidance documents on a variety of topics. However, hospitals may elect not to accept Medicare and Medicaid patients and, thus, are not compelled to comply with CMS guidelines. Most hospitals do accept these two federal insurance types and are CMS compliant.
- *Accrediting groups* are voluntary agencies that visit healthcare facilities to survey for compliance with standards of practice. The Joint Commission reviews the vast majority of hospitals and treatment centers. The Joint Commission has deemed status recognized by the CMS to verify that hospitals meet or exceed Medicare's requirements. Most healthcare facilities are accredited by the Joint Commission. However, some institutions are reviewed by other groups including Det Norske Veritas (DNV) and the Healthcare Facilities Accreditation Program (HFAP), which both have CMS-deemed status. In addition to CMS-deemed regulatory agencies, specific hospital departments may have additional regulatory agencies that oversee their work. One example is CLIA, which regulates laboratory testing. CLIA requires clinical laboratories to have certification by their state as well as the CMS before they can accept human samples for diagnostic testing.

- *State and local agencies* such as state-level departments of health.
- *National Fire Protection Association.* This trade group was formed before the turn of the 20th century by a group of insurance companies to minimize fire hazards and thus insurance losses. The NFPA has many codes, but two are associated specifically with healthcare and medical technology. In 1984, the group developed Health Care Facilities Code (NFPA 99), which is specifically associated with healthcare facilities. These guidelines are designed to "minimize the hazards of fire, explosion, and electricity in health care facilities providing services to human beings" (NFPA 99, 1984). Life Safety Code (NFPA 101) is a broad document that establishes construction standards for factories, hospitals, stores, and other structures, addressing emergency exits, fire protection equipment, and building materials. Some states have adopted NFPA 99 and NFPA 101 as state law.

Code Compliance: The Joint Commission and other accrediting bodies establish standards, policies, and procedures to guide review of healthcare facilities, with the aim to measure and improve performance in the delivery of safe and effective patient care. For example, the Joint Commission divides facility reviews into standards such as the Environment of Care (EC) and Information Management (IM). Each standard has many parts that set performance expectations and goals (Elements of Performance) for a wide variety of activities. In addition, the Joint Commission establishes yearly National Patient Safety Goals (NPSGs). These broadly defined objectives are designed to improve patient care and are revised regularly to reflect emerging trends or concerns. NPSGs are incorporated into the Joint Commission standards.

A team from the accrediting body will usually visit facilities unannounced. The survey conducted often utilizes a technique to assess performance that evaluates the interactions between a selected high-risk patient and facility from admission to discharge. This "tracer methodology" is designed to analyze care, treatment, and services using actual patients to assess compliance with standards. Survey findings result in an accreditation action that, at best, accredits the facility for 3 years.

Clinical engineers are tightly woven into the accreditation process, with focus on the performance expectations associated with medical technology, especially related to completeness of the device inventory and to device support activities such as preventative maintenance (PM) and recall management.

Standards are recommended best practice guidelines developed through discipline-specific stakeholder consensus. Standards do not require legal compliance, but professions or other groups may require compliance. A standard is designed to ensure consistent product or process outcomes to meet client needs. Similar to a regulation, organizations create a standard and seek feedback from the public. The American National Standards Institute (ANSI) and the International Organization for Standardization (ISO) are umbrella groups that recognize and catalog voluntary consensus standards. The Association for the Advancement of Medical Instrumentation (AAMI) is one organization that develops and publishes standards for advancing the safe and effective use of medical technology. Clinical sites follow recognized standards to offer the best possible patient care with the least risk. Some examples include:

- *AAMI EQ89, Guidance for the use of medical equipment maintenance strategies and procedures.*
- *AAMI EQ56, Recommended practice for a medical equipment management program.*
- *ANSI/AAMI 60601-1, Medical electrical equipment, Part 1.*
- *ANSI/AAMI/IEC 80001-1, Application of risk management for IT networks incorporating medical devices.*

Technical Information Reports (TIRs), are guidance documents that can help groups interpret and implement standards. TIRs can serve as practical guides to promote compliance with a standard. Some examples of TIRs include:

- *AAMI TIR57, Principles for medical device security—Risk management.*
- *AAMI TIR69, Risk management of radio-frequency wireless coexistence for medical devices and systems.*

Government regulations

On May 28, 1976, the federal government provided the FDA the responsibility and authority to regulate medical devices under the *Medical Device Regulation Act* or *Medical Device Amendments of 1976*. At the time, technology used in patient care was emerging as a tool, with just a few categories of devices available. However, healthcare technology was rapidly embraced and incorporated into general patient care. For example, infusion pumps were utilized to deliver fluids to almost every patient, in contrast to gravity systems (requiring no technology) utilized just a few years earlier.

In response to societal concerns over safety, the federal government passed the *Safe Medical Devices Act of 1990* (SMDA). This act established the three classes of medical devices. In addition, it established the requirement for pre-market approval by the FDA. It also required hospitals to identify and report serious problems with medical devices to the FDA. This act has been amended several times as technology has evolved and regulations have changed. Most of the changes to this act affect medical device manufacturers.

The *Health Insurance Portability and Accountability Act* of 1996 (HIPPA) regulates patient privacy. Medical devices are impacted by HIPPAA because patient information can be stored on many types of technology. Thus, HTM professionals work to ensure facility and technology compliance with privacy and security of patient information.

Regulations and codes evolve and change as problems arise or technology expands into new areas. Some laws evolve at a slower than expected pace and may result in challenging compliance when the regulations are unclear in a new area or antiquated as healthcare expands.

Medical device safety

A variety of groups ensure medical device safety with oversight associated with the stage of development and deployment. The FDA has a well-defined process for pre-market approval for Class III devices, including most of the technology associated with clinical engineering. The FDA pre-market approval review is shaped by federal regulations associated with federal law. Some clinical engineers are involved in the design, development, and review of new devices, especially those who work directly for the medical device manufacturers. However, most hospital-based clinical engineers have little influence over new devices that are acquired by a healthcare institution. While a great deal of resource material is available that is associated with the pre-market process, the focus of this section will be on the post-market regulations and codes that form a foundation for clinical engineers as they work to ensure safe and effective patient care.

Medical device safety concerns emerged in the 1970s as technology was introduced more commonly in patient care. Little regulatory oversight existed until 1984 with the creation of NFPA 99. As mentioned earlier, NFPA 99 is a tool that sought to minimize harm caused by fire, explosion, and electrical hazards. Thus the cornerstone of NFPA 99 standards is related to electrical systems and performance.

NFPA 99 requirements are a fundamental body of knowledge for HTM professionals. Many states have established this code as state law, thus compliance is mandatory. NFPA 99 is revised regularly, and adjustments are communicated throughout the HTM community. The 2012 version will be referenced in this text.

In offering a review of NFPA 99 components, the content is not meant to be exhaustive but instead offer a brief overview to help clinical engineers understand the overarching focus of the code. Additional research, investigation, and understanding will be needed to fully support the code and promote the safe use of devices.

Some NFPA 99 definitions include:

- *Patient care vicinity.* The space around patients where examinations or treatments occur is defined to be 6 feet around the bed and 7 feet 6 inches above the floor. The patient care vicinity space does not move with the patient (e.g., when a patient is walking in a hallway). In addition, this space does not extend beyond walls that may be present within the 6-foot space around the patient bed. The patient care area is associated with a variety of code sections that are different than non-patient care areas such as waiting rooms, nurses' stations, and so on.
- *Patient connection.* This is defined as an intended connection between a device and a patient that can carry electrical current. This can be an invasive connection such as an implanted wire or an electrical connection such as an ECG electrode.
- *Anesthetizing location.* Clinical care areas that are utilized to deliver anesthetic agents are subject to extensive regulations because of increased patient vulnerabilities in these spaces. Although several types of anesthesia exist (local and regional), the NFPA definition is associated with the application of general anesthesia (generally incorporating loss of consciousness and mobility). Generally, the clinical care area is the operating room.
- *Wet location.* This patient care area is associated with wet conditions (other than routine housekeeping, toilets, and washbasins). Wet locations have a variety of additional regulations to promote patient safety. The 2012 edition of NFPA 99 established all operating rooms as wet locations, thus requiring special protections against shock. However, the code does permit a risk assessment to determine whether an individual operating room is actually a wet location based

on the types of surgery performed. Guidance on the risk assessment process is available from the American Society for Health Care Engineering (ASHE). The ASHE Risk Assessment Tool is available from the ASHE website.

- *Patient-owned equipment.* Patients often bring personal devices with associated power cords into the patient care area. For example, many patients want to utilize their own insulin pump or CPAP machine. NFPA 99 10.4.2.1 does permit the use of patient-owned devices in the patient care vicinity, even those that do not comply with NFPA requirements (e.g., a ground cord), as long as the patient care staff conducts a visual inspection for "proper working order" or "worn condition." Many hospitals clarify the definition and use of patient-owned equipment to ensure patient safety.
- *Fixed equipment.* These are devices that are permanently hardwired to the hospital electrical system (e.g., imaging devices or sterilizers).
- *Portable equipment.* These are devices that have a power cord. While the devices may be mobile, some technology governed by the portable equipment NFPA requirements might never move from their initial installation location.

NFPA 99 dictates many electrical characteristics of a facility. For example, code 6.3.2.2.1.1 requires that the ground pin of an electrical receptacle be located at the "top" of an outlet. Most residences and businesses do not use this orientation. The purpose of this orientation has two components. First, should a power cord be pulled out by the cord rather than close to the receptacle, the last pin in contact with the source would most likely be the ground pin because most receptacles are lower to the floor than a person's hand. A second benefit is associated with the ability to short the neutral and hot pins by a falling object, should a power cord be slightly unplugged from the receptacle. For example, imagine a falling thin metal tray behind a cabinet or bed. A gap could be present between the power cord and the receptacle. However, the ground pin placement at the highest point would prevent a connection between the hot and neutral pins.

The National Electric Code (NFPA 70) is related to NFPA 99 because both require hospital receptacles in patient care areas to be hospital grade. The construction of these receptacles offers more resilience associated with grounding, reliability, assembly integrity, strength, and durability. Underwriters Laboratories has established tests that help ensure safe use in rough or high-abuse environments. Hospital-grade outlets have a green dot on the face to indicate compliance with these performance requirements. An example of a design specification is related to the retention force a receptacle must exert on a ground pin of a power cord. NFPA 99 6.3.3.2.4 requires a minimum force of 4 ounces. This means that a 4-ounce weight hung from a ground pin must be retained. Note that NFPA 99 10.2.2.1.1 requires power cords to have a ground conductor. These receptacle standards have minimized the yearly evaluations of electrical receptacles that were required before 1996.

Emergency power in times of electrical power outage is a grave concern for patient care, with special focus on life-saving devices such as ventilators. NFPA 99 6.4.3.1 requires power restoration within 10 seconds for receptacles that are colored red or have faceplates labeled "emergency power."

Many sections of NFPA 99 specify electrical safety requirements associated with medical devices. For example, NFPA 99 10.3.2.1 requires that the electrical resistance measurement between any conductive surface of a device to the ground pin of the power cord be no more than 0.5 ohm. As mentioned in Chapter 2, Healthcare technology basics, four currents are generally associated with risk to the user and/or patient, and these are regulated by NFPA 99 depending on device type (fixed or portable). For example, NFPA 99 10.3.5.1 establishes that the earth leakage current (current that flows between the power supply of the device through the ground of the power cord) is limited to 300 μA for portable equipment. Touch leakage current (current that flows between a person if he or she touches any part of the device and electrical/earth ground) is limited to 100 μA when the ground connection is intact. Patient leakage current (current that flows between patient connections and electrical/earth ground) is also limited to 100 μA. NFPA has adjusted these limits as the code has evolved since 1984. Technicians must be aware of the most recent requirements applicable to the healthcare facility.

Operator and user manuals, as mentioned in Chapter 2, Healthcare technology basics, are required to be available, stipulated in NFPA 99 10.5.3.1.2. The code identifies specific content that must be included in the manuals, including step-by-step PM procedures as well as schematics. In some instances, medical device manufacturers are reluctant to provide detailed service manuals. However, clinical engineers have utilized this section of the code to obtain detailed manuals to maintain device service and support.

As mentioned earlier, NFPA 101 establishes construction standards for the protection of occupants in many types of facilities, including factories and stores. Chapters 18 and 19 of the code specifically address healthcare environments. The code stipulates means of egress, fire protection and control, fire alarm and drill requirements, and construction constraints. Most NFPA 101 requirements are associated with the design and construction of facilities, as well as fire safety. Clinical engineers typically need to focus on just a few sections of the code as part of the hospital emergency response planning team. Regulations associated with hallway usage and door locks may be an area of interest.

Fire safety

Fire safety in healthcare facilities has improved dramatically but remains a concern because patients are typically unable to evacuate in an emergency. Additionally, normal processes within the hospitals can create high risk for starting a fire. One such environment is the operating room, where electrosurgical equipment may come in contact with flammable items, such as paper drapes placed over a patient or combustible anesthetizing gas. The confluence of these two items is generally the cause of most operating room fires. The code promotes a compartmentalization approach and encourages fire containment through construction techniques and barriers (e.g., doors, fire-rated walls) and limitation of potential spread through penetrating spaces (e.g., elevator shafts or conduit shafts for cables and pipes). Fire drills and emergency preparedness activities are stipulated in NFPA 101. These healthcare facility requirements are generally managed by the physical plant, although clinical engineers often are included as a team member, bringing knowledge of the technology utilized in the patient care areas. Clinical engineers may also participate in investigation activities after a fire event, especially if medical equipment was involved.

Radiation safety

Ionizing radiation is utilized in both diagnostic and therapeutic devices. As an example, x-rays utilized in imaging devices are very common. Ionizing radiation can be found in imaging suites, nuclear medicine departments, and cardiac catheterization spaces. Safe utilization for both patients and staff is critical to avoid injuries such as burns and cancer. Safety begins with a complete inventory of all devices that incorporate radiation. These devices have additional regulations and should receive special attention associated with their PM activities to keep devices in safe working condition. Outside service support may be needed when technical specialization is warranted.

HTM professionals who work in areas with radiation-related devices are required to be monitored for exposure through the use of badges that can track radiation. Badges utilize several techniques to measure exposure but are generally evaluated once per month. Several organizations, including the Nuclear Regulatory Commission (NRC) and the Occupational Safety and Health Administration (OSHA), regulate the environment. For example, the NRC stipulates that an institution must have a radiation safety officer. Both the NRC and OSHA set limits for workers related to radiation exposure. The Center for Devices and Radiological Health (CDRH), a division of the FDA, is associated with safety requirements for medical device manufacturers in controlling radiation-utilizing technology.

Radiation safety is promoted by shielding, limiting the amount of time in the environment, and increasing the distance from radiation sources. Shielding may include gloves, aprons, and partitions. Excellent communication with the users of this type of medical technology is also critical to promote safety. The quick identification of device malfunctions is achieved when solid rapport between staff and HTM professionals exists. Even when in-house service is not warranted, communication with the hospital HTM department can encourage a relationship that successfully manages the human side of technical repairs while tracking causes and seeking to identify patterns that could be utilized to mitigate equipment malfunction.

Many ionizing radiation—related devices are also of high value to the healthcare facility, both in acquisition cost and in revenue generation. Malfunction can result in both safety concerns and financial implications when technology is unavailable. Balancing the competing demands of utilization and safety is a hallmark responsibility of the clinical engineer.

LASER safety

The acronym *LASER* stands for *light amplification by stimulated emission of radiation*. As discussed in Chapter 2, Healthcare technology basics, this device uses specially treated light to alter tissues in a specific way. LASERs typically coagulate or vaporize (cut) tissue.

In general, LASER light waves have the following characteristics:

- *Coherent*. All light waves are in a single phase.
- *Monochromatic*. All light waves are of a single color (frequency).
- *Collimated*. All light waves are capable of staying together in a tight beam over long distances.

Some LASERs are capable of producing more than one color and produce different tissue effects. LASER light can be made from four different sources (mediums): solid, liquid, gas, or electronic.

Some common types of medical LASERs are:

- CO_2 (carbon dioxide; a very common surgical laser for cutting and coagulation);
- Nd:YAG (neodymium yttrium aluminum garnet);
- tunable dye;
- argon; and
- ruby.

Some functions of LASERs include corrective eye surgery, removal of tumors, general cautery during surgery, removal of tattoos, and skin resurfacing. As a result, LASER-related devices are found in many areas of a healthcare facility. The common use of CO_2 LASERS in operating room spaces establishes fire as a concern because the LASER may serve as a source of ignition.

LASERs are divided into classes that identify their risk to humans, with burns and eye injury the most common concerns. Classes 3 and 4 LASERs are most commonly found in healthcare facilities and present special safety requirements including eye protection, signage, and training. As with ionizing radiation technology management, LASER devices require an accurate inventory, tracking of PM and repair records, and appropriate training. A LASER safety officer may be responsible for these activities in close collaboration with clinical engineers. The Joint Commission and OSHA jointly prepared a safety guide entitled *LASER Beam Safety Scheme* (the Joint Commission, 2015).

Cybersafety

Risk management around cyberthreats, vulnerabilities, and mitigation associated with medical devices is emerging as a critical and separate area. Many well-known healthcare computer system assaults have been related to medical record data. Patient health data compromises are especially important as the information cannot be "refreshed" after an attack as a credit record can be. Identify theft and medical fraud related to patient-identifying information is deeply worrisome. In contrast, few device-related cyberattacks have been launched.

Although violations of HIPAA requirements and privacy leaks were initially the concern associated with cybersecurity, as threats have increased, patient safety has emerged as a growing concern. As more devices are connected to each other and to the Internet, control over technology to change device modality or performance is a growing concern.

In 2016, AAMI published TIR57, *Principles for medical device security—Risk management*, which offers insight into medical device security management. More information regarding cybersecurity is described in Chapter 5, Information technology.

Infection control

The healthcare environment is composed of a variety of microorganisms that can transmit disease to patients, staff, and visitors. Vehicles for this transmission include the air, medical equipment, as well as from person-to-person. Staff handwashing expectations are often communicated to the public. For example, the general population likely can describe the contamination associated with the neckties of doctors. However, the processing of devices and instruments to render them not just clean but contaminant free has emerged as an important part of the clinical engineering scope of practice. AAMI has published *Checklists for Preventing Healthcare-Associated Infections*, a free guide that may serve as a reference for appropriate actions by HTM professionals.

Hospital-acquired infections, also termed *nosocomial infections*, are related to patient stays in medical facilities and are an increasing concern to both regulatory bodies and medical insurance carriers. A patient who acquires a disease as a result of a course of treatment will spend a significantly longer time in the hospital and may have impactful health implications including death. Thus, reducing transmission of disease through device cleaning, de-contamination, disinfection, and sterilization are critical activities.

HTM professionals play an integral part in the prevention of the spread of disease associated with medical technology de-contamination processes. They work to avoid the use of so-called "dirty equipment," devices that harbor bacteria and other contaminants. All medical devices, such as infusion pumps, must be de-contaminated in a specific and detailed way to minimize the spread of disease to patients, staff, and visitors. Manufacturers must provide descriptive practices to characterize the method to de-contaminate equipment, called *reprocessing*, and these practices should be detailed in the instructions for use (IFU). The sterile processing or central supply department (name usage varies) is tasked with these activities that are typically intensely physical and require vigilant attention to detail.

Clinical engineers are often engaged to work with sterile processing staff to ensure that IFU instructions are implemented as expected and that results are satisfactory.

A confounding factor in infection control is the increase in complexity of medical devices. These devices offer more electronics and tiny physical qualities such as tubes or pathways. In addition, the disinfection of a new device is unlikely to be a high priority during design. Thus, the skill set that is required to process these complex devices can exceed the expectations of sterile processing technicians, whose pay levels have historically been low, which results in turnover and reliability concerns. Reducing infection transmission requires training, workload analysis, and the collaboration of clinical engineers.

Surgical instruments and endoscopes are uniquely of concern, and thus standards and guidelines have been created to address the unique reprocessing demands of these devices. In addition, the FDA offered a safety communication in 2019 regarding the cleaning of duodenoscopes in response to continued de-contamination challenges associated with these instruments (FDA, 2019). Related standards include:

- *ANSI/AAMI ST90, Processing of health care products—Quality management systems for processing in health care facilities.*
- *ANSI/AAMI ST91, Comprehensive guide to flexible and semi-rigid endoscope processing in health care facilities.*
- *ANSI/AAMI ST79, Comprehensive guide to steam sterilization and sterility assurance in health care facilities.*

Research activities and the Institutional Review Board

Researchers may use data collected from medical devices and information gleaned from patient records to identify issues with these devices and optimize healthcare. De-identified data can be used in many ways to develop decision support algorithms or identify best practices on how to treat specific patient populations. However, the process to access the data for research purposes is carefully regulated to protect the privacy of the data sources.

Research activities in many disciplines are governed by an Institutional Review Board (IRB), established by each institution under federal requirements. This administrative body is established by research organizations and is responsible for protecting the rights and privacy of human research subjects who consent to participate in research or whose information is utilized in studies. The current system of protection for human research subjects was influenced by the *Belmont Report*, written in 1979 by the National Commission for the Protection of Human Subjects of Biomedical and Behavioral Research (HHS, 1979). Regulations were established in response to unethical research projects conducted without oversight. The IRB has the authority to review, approve/disapprove, monitor, and modify all research activities as specified by federal regulation. Under HHS, the Office for Human Research Protections governs regulations for the protection of human subjects under the Code of Federal Regulations 45 CFR 46 (HHS, 2017). This regulation also includes four subparts; 45 CFR 46 Subpart A, known as the "Common Rule" was updated in 2018 and implemented effective January 21, 2019 (HHS, 2019). This subpart provides basic provisions for the IRB, informed consent, and assurances of compliance. Subpart B includes additional protections for pregnant women, fetuses, and neonates. Subpart C provides additional protections for prisoners, and subpart D provides additional protections for children.

Organizations and researchers must be careful as some quality improvement activities may be subject to human research protections. While quality improvement activities such as analyzing non-identifiable data or measuring and reporting provider performance data for clinical, practical, or administrative use may not satisfy the definition of research under 45 CFR 46.102(d), other quality improvement projects may meet the requirements. For example, a project involving an untested clinical intervention for purposes that include not only improving the quality of care but also collecting data on patient outcomes may constitute non-exempt human subjects research (HHS, 2017).

Medical device incidents

Despite the best efforts of the entire healthcare team, adverse incidents related to medical devices occur. Patients or staff may be harmed in some way. Injuries include electrical shocks, tissue damage, or delay in therapeutic care. Medical staff often focus on the physiologic impact of the malfunction, but the clinical engineer must establish whether the device or its system failed.

Accidents may be attributed to malfunctioning equipment, human error resulting in improper use, fundamental device design, or a lack of training. An additional possible cause may be associated with purposeful errors associated with anger, criminal intentions, or mental defect. Clinical engineers have an ethical responsibility to analyze adverse events to identify the cause and work toward mitigating the potential for a similar event in the future. With this requirement, clinical engineers must utilize a myriad of investigational tools and knowledge to better understand the fundamental causes of adverse incidents.

As part of the healthcare team, clinical engineers must take an active role in partnering with a facility risk management office to establish policies and procedures that will guide investigations in ways that offer the HTM department staff the opportunity to collaborate. Risk managers generally have the responsibility and expertise to lead an investigation after an adverse event. However, risk managers may lack significant knowledge of medical devices and the systems in which they operate. Thus, clinical engineers are critical partners with risk managers in establishing policy to guide activities immediately after an incident. These policies should identify responsible staff, determine the process to protect evidence, and establish a documentation schema that is uniform and comprehensive.

Investigation

When technology is associated with an adverse event, clinical engineers often gather evidence and analyze observations to identify the cause or causes of the incident. Conclusions must be adequately justified and reduce risk in the future. Although ethical responsibilities serve as a foundation, legal liability is a constant concern and will shape investigatory actions. When an event is serious, a lawsuit is likely and may affect the investigation. Liability insurance is held by multiple parties including the medical practitioners as well as the institution and is very expensive. Litigation can result in significant financial awards. Thus, investigations should be driven by a pre-established framework and produce solid results. However, pending litigation may hamper communication or be counterproductive.

In addition to malpractice litigation concerns, laws and regulatory guidelines affect investigations. Federal laws such as the Medical Device Amendments of 1976 and the SMDA establish incident response requirements. Facility accreditation through bodies such as the Joint Commission could be jeopardized. With this backdrop, clinical engineers should seek to utilize a large list of investigatory tools such as those described in detail by Dyro and by Hyndman and Shepherd (Dyro, 2004; Hyndman & Shepherd, 2015). Suggestions included items like hand tools, a flashlight, and a camera. However, as technology has evolved and devices are integrated into more complex systems, critical tools might include network analyzers and cable testers. Clinical engineers should think about the tools that may be needed to evaluate system performance errors in addition to device malfunctions before there is a need to conduct an investigation, ideally at the time of technology purchase.

Before an adverse event, the HTM department staff should be well-trained to respond appropriately and in ways that support the identification of the cause of the incident. When an incident occurs, preservation of the evidence is critical. Actions should include isolating the device and all ancillary components (including disposables) in a secure location, archiving electronic information, and documenting the event in as much detail as possible. The response protocol should dictate that photographs are taken immediately after the event. User experiences, in detail, can be critical in the investigation. For example, the delay over a weekend may diminish the quantity of critical information. Immediately post-event, testing of the device or components may mitigate faults and destroy crucial evidence.

Once a thorough investigation has begun, the testing of technology must be purposeful and thoughtful as actions could erase memory, change settings, or adjust performance. In addition, the appropriate tools, including device and system analyzers, must be available. Often time is of the essence in response to an adverse event. However, attention to detail and careful evaluation of unintended consequences is important. In addition, clinical engineers may seek the advice of experts associated with both the technology as well as the type of event. In addition, the assistance of device manufacturers may be vital as performance analysis may require significant technical expertise beyond an in-house HTM department.

Utilizing the extensive work of Shepherd in the creation of system risk models, four broad components of the technology mini-system have been identified as the possible categories of failure: the *device*, the *operator*, the *patient care area/facility*, and the *patient* (Hyndman & Shepherd, 2015). These are described in Chapter 3: Healthcare technology management. The device itself can contribute to an adverse event through component malfunction, deterioration, improper maintenance, or the human factors design. The operator of the device may have gaps in training or may experience diverted attention caused by fatigue or stress. The patient care area may experience utility malfunctions or changes in environmental conditions such as temperature. In addition, the interconnectedness of devices may experience a fault in one component that produces negative impact elsewhere. Finally, the patient may contribute to an adverse event both actively by violating medical orders or passively as a result of physical condition. Both the user and the patient could promote an adverse event when harm is intended. These categories and examples are not intended to be exhaustive. The highly complex systems of medical technology result in a myriad of failure opportunities that extends far beyond the reach of this text.

Clinical engineers should gather evidence in all four broad categories by looking for records and documents that reflect clinical notes, maintenance, recalls, training, and device standards. Broad investigation can avoid a natural tendency to either focus on user error or device malfunction. In addition to documents, interviews can be a vital resource to better understand an adverse event. Investigators should be sure to interview all people who may have relevant facts and utilize open-ended questions that do not guide the subject to respond in a narrow way. Careful notes or recordings can be helpful when conducting interviews.

After the investigation, a final report will include a description of the incident including the device and accessories, an analysis of the device and facility testing, summaries of interviews, and possible user errors that may have contributed. The report should also include a reconstruction of the event (based on evidence) and recommendations that include actions and activities that could mitigate future related events. When an event is complex or highly technical, a summary may be useful to better convey the information.

Root cause analysis

Root cause analysis (RCA) is a tool used to analyze events. The method can promote the identification of underlying problems and includes both interface errors as well as hidden challenges. RCA is commonly used and was detailed in a 2016 Joint Commission publication, *Root Cause Analysis in Health Care: Tools and Techniques* (the Joint Commission, 2016). A team examination of the events that occurred before an event should be able to identify the causes of the event. RCA seeks to avoid placing "blame" on individual actions but instead looks to understand and improve the system that resulted in incorrect actions. Because the method is focused on the broad components of the patient care environment, RCA may be termed or viewed as *systems analysis*.

The CMS has prepared a guidance document to assist in the utilization of RCA (CMS, n.d.). As described in the document, RCA examines the contributing factors associated with the adverse event with a prompt "What was going on at this point in time that increased the likelihood the event would occur?" (p. 5). Utilizing this approach, investigators are more likely to look beyond the device, the user, and the patient to identify issues such as staffing levels, "work-arounds," and communication barriers.

RCA also differentiates between contributing causes and root causes that are fundamental. Most root causes do not have short-term resolutions. For example, if missed PM checks on a device is a contributing cause to a device malfunction, an examination of the causes of missed PM checks is more closely associated with root cause identification. Fundamental root causes of technology-related unintended harm to patients are becoming more difficult to identify within highly complex and integrated systems. Multidisciplinary teams incorporating risk managers, clinicians, engineers, and IT professionals are critical to dissecting an event to understand the fundamental causes.

Finally, RCA seeks to design corrective actions that are feasible, reasonable, and implementable. Changing processes, systems, and core practices can reduce the likelihood of a similar event in the future.

Reporting requirements

Sharing information among facilities when adverse events occur is critical in the detection of trends or patterns and can improve the safe use of technology in the future. The SMDA mandated reporting of specific events to the FDA. MedWatch is the FDA medical device safety program. Reporting requirements are shown in Table 4.1 (FDA, n.d.):

Note that the federal 10-day reporting window often creates a sense of urgency after an incident, driving quick timelines for investigations. Reports are archived in a searchable database that can be a tremendous resource to clinical engineers. The Manufacturer and User Facility Device Experience (MAUDE) database can be found on the FDA website (www.fda.gov) utilizing a search for MAUDE. Exploring this database can be exceptionally informative as it offers insight into common sources of fault. One can try searching for a common device such a ventilator and read through the reports.

Table 4.1 Reporting requirements for the FDA medical device safety program, MedWatch.

Reporter	What to report	Report form #	To whom	When
User facility	Device-related death	Form FDA 3500A	FDA & manufacturer	Within 10 work days of becoming aware
User facility	Device-related serious injury	Form FDA 3500A	Manufacturer.FDA only if manufacturer unknown	Within 10 work days of becoming aware
User facility	Annual summary of death & serious injury reports	Form FDA 3419	FDA	January 1 for the preceding year

More information about medical device incident reporting can be found on the FDA website utilizing a search for medical device safety. Clarification and definitions for healthcare facilities are provided in the FDA publication *Medical Device Reporting for User Facilities* (Lowe & Scott, 1996).

In addition to the FDA, some states require reporting to the US Department of Health. Accrediting bodies such as the Joint Commission may also require reporting of adverse incidents. ECRI (formerly the Emergency Care Research Institute) is a private organization that collects medical device problem reports voluntarily submitted from their website (ECRI Institute, n.d.). ECRI then investigates and compiles these reports to prepare yearly hazard reports as part of its monthly publication *Health Devices.*

Clinical engineers should value the shared information associated with adverse events at other facilities, utilizing the information to better protect the patients and staff. Proactive interventions (e.g., improved training or signage) can reduce the likelihood of the need for an investigation in the future.

Recall management

When issues are identified by manufacturers or the FDA, medical devices may become subject to a recall. This action does not necessarily require that a device be taken out of service but in fact requires some corrective action to ensure safe and effective performance. The FDA issues a voluntary recall to protect public health from products that present a risk for injury. A recall is a voluntary action that takes place because manufacturers and distributors carry out their responsibility to protect the public health and well-being from products that present a risk for injury or gross deception or are otherwise defective. Corrective actions may include device checks, adjustments, updates, or repairs. Software updates or patches may also be associated with recalls.

Recalls are categorized by the FDA as Classes I, II, or III. Class I recalls are the highest priority action because the potential for serious harm, such as death, is probable. Class II recalls reflect a concern that temporary adverse health consequences could occur. Class III recalls indicate that corrective action is needed but that adverse consequences to patients or staff are unlikely (FDA, n.d.). Note that this classification system is in reverse to the medical classifications described in Chapter 2, Healthcare technology basics, in which Class I medical devices have the lowest risk and Class III devices have the highest risk.

Notification of recalls can occur in a wide variety of ways through public posting on websites by both the manufacturer as well as the FDA. Press releases may reach public news outlets. In addition, subscription-based services such as ECRI notify healthcare facilities of recalls. Email and postal services are also utilized to communicate recalls. Challenging to the dissemination of recalls is channeling the needed information to the responsible professionals. A communication plan can be critical in the notification and mitigation of device hazards.

Recall trends have increased exponentially in number as devices become more numerous and more complex, especially with the heightened risk for cybersecurity vulnerabilities. With this shift, workloads associated with the management of recalls has become extensive and heavy. When recalls are associated with a large number of devices, the manpower required to carry out the mitigating actions can stress HTM department staff. When patches are released to mitigate cyber-risks, hospitals may have to wait months between the notification of a recall and the release of the software. This time delay makes it a challenge to track and complete all of the recall actions. In addition, this can affect patient care if devices are removed from use during the needed corrective activities.

Tracking recalls involves identifying the involved devices owned by the facility as well as locating these devices. Hospital inventory databases must be robust to readily associate the recalled devices with hospital-owned equipment. Work orders must be created and completed for each of the corrective actions identified in the recall. In addition, an HTM professional is often tasked with ensuring that all affected devices are found and appropriate actions are taken before deeming the recall complete.

AAMI provides a list of nine best practices associated with recall management (Caceres, 2014):

1. Have a written recall management policy and inventory.
2. Always know what inventory is on hand.

3. Ensure there are processes in place to receive recall notifications.
4. Work with a multidisciplinary team to manage and prioritize recalls.
5. Stay in contact with end users of products to find out about recall notices received, required actions, and next steps.
6. Try innovative ways to track down equipment.
7. Document everything that occurs related to both equipment inventory and equipment recall.
8. Communicate to leadership what resources are needed to effectively manage recalls.
9. Consider how the organization will dispose of recalled products.

Recall management depends on clear communication and efficient tracking of corrective actions. Building adaptability into the HTM workload is critical to manage unplanned recall notices that affect a large number of devices or critically needed technology.

Risk management

Medical treatment and overall care manage a challenging and delicate balance between benefits and risks. Patients and staff experience no risk if the patient does not visit the healthcare facility to receive treatment. However, the patient then receives no benefit (possible wellness). Thus, HTM professionals manage risk as best as possible while still providing effective patient care. Understanding the unique risks of healthcare is helpful in framing the balancing act.

People who seek medical treatment are subject to high risk for a variety of reasons that are unique to the healthcare environment. These factors include:

- Many devices, both in type and quantity, are utilized in patient care. The technology may compete for physical access to a patient and may interfere with other devices.
- The patient care environment can be harsh and unpredictable to technology. For example, sudden changes in patient condition may cause unplanned movement, significant exposure to fluids, or other technology-damaging actions.
- Patients are directly and electrically connected to technology. Within these connections are catheters and fluid-drenched environments that may promote unintended conduction of electricity. The skin, usually an excellent insulator, may be broken down, removing a layer of protection.

- Diverse healthcare providers may have limited understanding of a particular technology associated with a device or only a very basic understanding of physical principles such as electricity or fluid mechanics. Thus, predicting hazards may be challenging for clinicians.
- Technology is used by family members in diverse environments including patient homes. In some cases, the manufacturer did not expect or plan for the technology to be used in this environment.
- Patients may be immobile or unconscious, preventing the usual human protection reflexes such as moving away from a hazardous situation.
- Interconnectedness of medical devices can present unanticipated behavior or interactions.

Risk is often categorized into four simplistic groups:

- low impact, low probability;
- low impact, high probability;
- high impact, low probability; and
- high impact, high probability.

In healthcare, impact is directly related to patient or staff harm, that is, physical injury, on the continuum between inconvenience and death (not binary: low or high). In addition, risk likelihood may occur in varying degrees. The simplification of categories can help clinical engineers discern truly crucial environments with high impact (life-sustaining devices) and high probability (highly complex or prone to errors).

With these categories in mind, HTM professionals must explore risk and risk assessment to best ensure patient safety. To guide this exploration, AAMI published a guide in 2015 titled *Making Risk Management Everybody's Business: Priority Issues from the 2015 AAMI/FDA Risk Management Summit*... available at this link http://s3.amazonaws.com/rdcms-aami/files/production/public/FileDownloads/Summits/2015_Risk_Mgmnt_Summit_Report.pdf?/Risk_Management_Summit_Report. This document provides clinical engineers tools to manage the risks associated with the rapid innovation and adoption of technology when clashing with human culture. In this document, risk management was defined as the "systematic application of management policies, procedures and practices to the tasks of analyzing, evaluating, controlling, and monitoring risk" (p. 11). This definition should drive clinical engineering actions:

- Analyze risk;
- evaluate risk;
- monitor risk; and
- control risk.

Once these tasks are begun, policies, procedures, and practices to mitigate risk in ways that clinicians can understand and embrace should be generated. This requires diminished isolation among clinicians, risk managers, and technology experts. Communication and coordination among groups is vital in the effort to minimize risk. Broadly, ISO 31000:2018, *Risk management—Guidelines*, provides general principles for managing risk. The more focused standard ANSI/AAMI/ISO 14971:2019, *Medical devices—Application of risk management to medical devices*, may offer additional guidance in risk management for HTM professionals.

Quality management practices

In CFR 21, the FDA offers the HTM profession definitions of *quality*, a *quality policy*, and a *quality system*.

- *Quality* means the totality of features and characteristics that bear on the ability of a device to satisfy fitness-for-use, including safety and performance.
- *Quality policy* means the overall intentions and direction of an organization with respect to quality, as established by management with executive responsibility.
- *Quality system* means the organizational structure, responsibilities, procedures, processes, and resources for implementing quality management.

A quality management system (QMS) is a set of institutional policies, processes, and procedures that an organization establishes and implements to ensure that it consistently delivers quality products and/or services to its customers or clients. ISO 9001:2015, *Quality management systems—Requirements*, describes QMS components and is broad with application to many types of industries. Fundamentally, QMS roots are within the *plan-do-check-act* sequence of activities, integrating multiple reporting levels of an organization into the process to ensure quality delivery of products and/or services.

- *Plan*. Establish a plan to meet client needs.
- *Do*. Design and implement operational activities to meet client needs.
- *Check*. Gather evidence to determine if the activities are effective, that is, document whether or not client needs are met.
- *Act*. Implement process changes to promote continuous improvement.

The interrelationship among the system components is illustrated in Fig. 4.1 and described in ISO 9001.

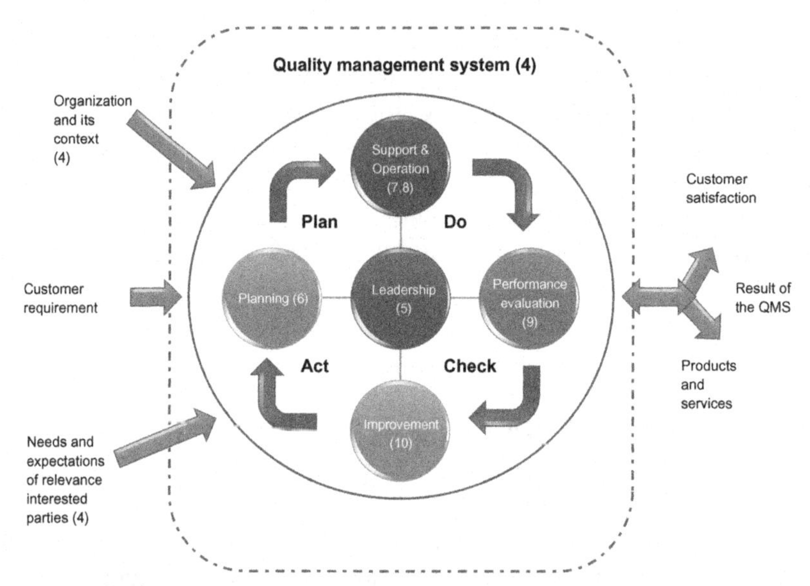

Figure 4.1 Quality management system components.

The diagram illustrates the role of leadership in the system, engaging decision makers in the quality assurance process. In addition, evidence-based decisions are critical for success. QMS requires an analysis of the smaller systems embedded within the larger systems—a common structure in healthcare. Medical facilities are often described as systems of thousands of interconnected but isolated systems. One contributing factor to this complex framework is the healthcare delivery system age. Hospitals were established in the middle to late 1800s. Thus, the environment has a long-established culture and hierarchy that may impede communication and collaboration.

ISO 9001 served as a framework to support more specific standards related to medical device manufacturers, such as ANSI/AAMI/ISO 13485:2016, *Medical devices—Quality management systems—Requirements for regulatory purposes*. However, the HTM profession does not have a QMS-related standard. Many HTM professionals argue that the components of QMS are integrated into professional practice, but under other names. Because QMS best practices are interwoven into accreditation standards, such as those by the Joint Commission and DNV, quality expectations and activities are in place in the healthcare environment. Implementation of a formal QMS process could promote cross-disciplinary expertise sharing, consistency in documentation, and careful selection of metrics to document quality.

QMS principles expand evidence collection beyond traditional HTM-related metrics, such as PM completion rates (a common calculation of devices due for inspection during a month in comparison with the amount of inspections completed). In addition, a broader and more purposeful plan for institutional quality can end the "break-fix" mentality of some HTM professionals. Clinical engineers should seek out metrics that are meaningful to the institutional leadership and clinicians. The QMS process could document and improve quality delivery of services to constituents including clinicians, patients, family members, and hospital support services.

Abbreviations

ANSI	American National Standards Institute
ASHE	American Society for Health Care Engineering
CDRH	Center for Devices and Radiological Health
CMS	Centers for Medicare and Medicaid Services
CLIA	Clinical Laboratory Improvement Amendments of 1988
DNV	Det Norske Veritas
ECRI	Emergency Care Research Institute
EC	Environment of Care
HIPAA	Health Insurance Portability and Accountability Act
HTM	healthcare technology management
IM	Information Management
IRB	Institutional Review Board
IFU	instructions for use
ISO	International Organization for Standardization
MAUDE database	Manufacturer and User Facility Device Experience database
NFPA	National Fire Protection Association
NRC	Nuclear Regulatory Commission
OSHA	Occupational Safety and Health Administration
QMS	quality management system
RCA	root cause analysis
SMDA	Safe Medical Devices Act of 1990
TIR	Technical Information Report
HHS	US Department of Health and Human Services
FDA	US Food and Drug Administration

References

Caceres, V. (2014). Clearing the shelves: Insights on how to handle medical device recalls. *Biomedical Instrumentaton and Technology*, *48*(6), 414–422.

Centers for Medicare and Medicaid Services. (n.d.). *Guidance for performing root cause analysis (RCA)*. Retrieved from <https://www.cms.gov/medicare/provider-enrollment-and-certification/qapi/downloads/guidanceforrca.pdf>.

Dyro, J. (2004). Accident investigation. In J. Dyro (Ed.), *The clinical engineering handbook* (pp. 269–281). Burlington, MA: Elsevier Academic Press.

ECRI Institute. (n.d.). *Medical device problem reporting*. Retrieved from <http://www.mdsr.ecri.org/information/problem.aspx>.

Hyndman, B., & Shepherd, M. (2015). *Theory and practice of device incident investigations. AAMI,* A practicum for healthcare technology management (pp. 87–113). Arlington, VA: Association for the Advancement of Medical Instrumentation.

Lowe, N., & Scott, W. L. (1996). *Medical device reporting for user facilities*. Rockville, MD: US Department of Health and Human Services.

The Joint Commission. (2015). LASER beam safety scheme. *Environment of Care News, 18* (2), 7–9.

The Joint Commission. (2016). *Root cause analysis in health care: tools and techniques* (5th ed.). Oak Brook, IL: Joint Commission Resources.

US Department of Health and Human Services. (April 18, 1979) *The Belmont report: Ethical principles and guidelines for the protection of human subjects of research*. Retrieved from <https://www.hhs.gov/ohrp/regulations-and-policy/belmont-report/read-the-belmont-report/index.html>.

US Department of Health and Human Services. (January 19, 2017). *OHRP regulations*. Retrieved from <https://www.hhs.gov/ohrp/regulations-and-policy/regulations/index.html>.

US Department of Health and Human Services. (April 2, 2019). *Revised commom rule regulatory text*. Retrieved from <https://www.hhs.gov/ohrp/regulations-and-policy/regulations/revised-common-rule-regulatory-text/index.html>.

US Food and Drug Administration. (April 19, 2019). *The FDA continues to remind facilities of the importance of following duodenoscope reprocessing instructions: FDA Safety Communication*. Retrieved from <https://www.fda.gov/medical-devices/safety-communications/fda-continues-remind-facilities-importance-following-duodenoscope-reprocessing-instructions-fda>.

US Food and Drug Administration. (n.d.). *Medical device safety*. Retrieved from <https://www.fda.gov/medical-devices/medical-device-safety>.

US Food and Drug Administration. (n.d.). *Recalls, corrections and removals (medical devices)*. Retrieved from <https://www.fda.gov/medical-devices/postmarket-requirements-devices/recalls-corrections-and-removals-devices>.

CHAPTER 5

Information technology

The clinical engineering–information technology interface

Historically, medical devices were stand-alone and did not interact with, interface with, or transfer data among systems. However, with more and more manufacturers creating devices that can transmit and receive data from other systems to enhance function and workflow, many medical devices now include either embedded software platforms or other hardware or software systems that look very much like an information technology (IT) system. This transition has brought clinical engineering and IT closer and closer together.

This convergence of technology has brought two different teams together and has highlighted differences in culture and process between the two groups. Clinical engineering approaches all medical equipment using the technology life-cycle management approach (see Fig. 5.1) discussed in Chapter 2, Healthcare technology basics. This life-cycle approach has been generally a hands-on approach in which clinical engineering staff interact face-to-face with clinicians in their units to discuss and resolve issues. This hands-on service level is generally opposite to the approach IT uses.

IT departments normally use Information Technology Infrastructure Library (ITIL) practices for IT service management. These practices emphasize aligning IT services with the needs of the hospital. The ITIL lifecycle suite, 2011 edition, has five volumes that cover the following processes:

1. *ITIL Service Strategy* understands organizational objectives and customer needs and provides direction on clarification and prioritization on investment into services provided by IT (Cannon, 2011). This stage covers processes such as:
   a. strategy management for IT services;
   b. service portfolio management;
   c. financial management for IT services;
   d. demand management; and
   e. business relationship management.

Introduction to Clinical Engineering
DOI: https://doi.org/10.1016/B978-0-12-818103-4.00005-3

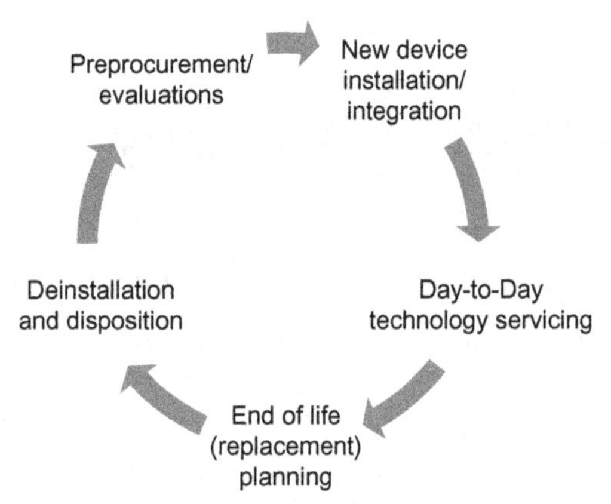

Figure 5.1 Phases of the technology support life cycle.

2. *ITIL Service Design* turns the service strategy created in the first volume into a plan for delivering the business objectives. It helps guide the design of services and processes to meet needs (Hunnebeck, 2011). This stage covers processes such as:
   - **a.** design coordination;
   - **b.** service catalog management;
   - **c.** service-level management;
   - **d.** availability management;
   - **e.** capacity management;
   - **f.** IT service continuity management;
   - **g.** security management; and
   - **h.** supplier management.
3. *ITIL Service Transition* develops and improves capabilities for introducing new services into supported environments (Rance, 2011). This stage covers processes such as:
   - **a.** transition planning and support;
   - **b.** change management;
   - **c.** service asset and configuration management;
   - **d.** release and deployment management;
   - **e.** service validation and testing;
   - **f.** change evaluation;
   - **g.** knowledge management

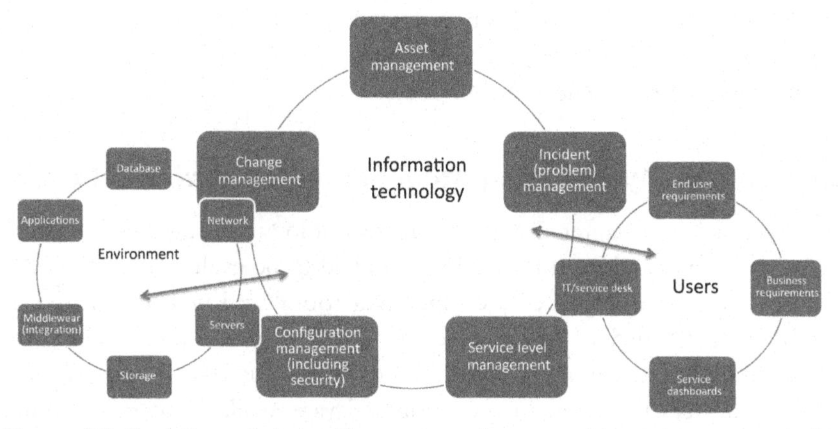

Figure 5.2 Depiction of how IT processes interact with end users and the environment.

4. *ITIL Service Operation* manages day-to-day operations in supported environments (Steinberg et al., 2011). This stage covers processes such as:
   a. event management;
   b. access management;
   c. request fulfillment;
   d. problem management;
   e. incident management
5. *ITIL Continual Service Improvement* focuses on incremental and large-scale improvements (Lloyd, 2011).

These processes within the service operation volume detail how IT interacts with its end users or clinicians. Generally, management of all these processes is done through a help desk or service desk that is reached through an email address or phone number. The help desk interaction is generally remote, where an analyst endeavors to solve the issue by using tools at his or her desk to resolve the request or issue. Fig. 5.2 depicts several of the IT processes and how they interact with both the end users through the service desk and the environment they manage.

The remote interaction IT utilizes to meet end-user needs is generally opposite to the direct interaction clinical engineering uses to resolve requests and incidents. This fundamental difference in service approaches may lead to cultural differences between the groups. These cultural differences can cause tension between the two teams when trying to create support plans for complicated clinical systems.

Interoperability and integration of medical device data

As described in Chapter 3, Healthcare technology management, historically, data generated from medical devices had to be evaluated at the location of the device (e.g., bedside, operating room) and was not logged in to the patient's medical record unless manually recorded by the clinician. As technology has evolved, medical device manufacturers have begun offering devices that contain ports and software available to deliver vital signs, images, alarms, and other data generated from the medical device to a myriad of other systems used in the hospital. This extension of Internet connectivity into physical devices is called the *Internet of Things (IoT)*. In this section, we will discuss integration of medical device data to the medical record and other systems. We will also discuss how these data are used in decision support algorithms, transmitted to secondary device systems, and used in other ways to support healthcare applications.

Integration to medical records

Hospitals and other healthcare organizations have many different computer systems used for everything including medical records, billing, and patient tracking. Electronic medical records are systematic collections of electronically stored health information that may include demographics, medical histories, allergies, medications, laboratory results, radiology images, and billing information. These records can be shared within a healthcare system using their enterprise-wide information system or across hospitals using exchanges. The electronic medical record system is one of many electronic systems hospitals use and is generally the repository for information coming from other systems. Many of these systems must communicate with, or interface to each other to acquire new information, retrieve information, or send out information. Most systems use Health Level Seven International (HL7) standards for transfer of data between software applications. HL7 standards focus on the application layer, which is layer 7 in the Open Systems Interconnection (OSI) model (Microsoft, 2017).

The foundation system required for all integration activities is the admission, discharge, and transfer (ADT) system. This system holds valuable patient information such as medical record number, age, name, and contact information. Using the ADT system, patient information can be shared, when appropriate, with other healthcare facilities and systems (McGonigle & Mastrian, 2012). This patient information is the backbone of device integration efforts as the data need to be transferred to the correct patient record.

There are two general workflows that can occur to transfer data from a medical device to a patient's medical record. One is a pull workflow in which data are pulled into the medical record from the device, and the other is a push workflow in which an order is pushed from the medical record to a device for a test or procedure.

Pull workflows

These workflows are generally used from within the medical record and are utilized primarily for vital signs capture. For example, physiologic monitors can send vital signs (e.g., heart rate, blood pressure) to the medical record automatically. This workflow requires two steps to be successful. The first step is the ADT association of the patient to the monitor, also known as *admitting the patient to the monitor.* This step is critical to ensure that the results from one specific monitor are aligned to the correct patient. Once the monitor is associated to the patient, a clinician goes into the medical record and "pulls" the vital signs on the monitor into the medical record. These vital signs show up in fields that are allocated to the values and identify the time at which the values were acquired. The nurse then validates that the values in the record match the values on the monitor.

This simple two-step nursing workflow requires quite a bit of setup, programming, and testing on the back end to ensure that the system seamlessly transmits data across the network. Some monitoring vendors have a separate network through which the physiologic monitors communicate. Connection of this network to the ADT system and medical record generally requires several servers and routers. Because this system requires significant testing, most larger hospitals and healthcare systems will not only set up a system used for real time, or production, but will also set up a test system or development system that mirrors the production system. This development system is used to test programming, software updates, and patches without affecting production. When all of the issues are worked out in the development system, the changes can be made to the production system. Additionally, larger healthcare systems that deem this system to be critical to workflow may also create a backup, or failover, system that is also live in the production environment.

Clinicians can switch to this backup system in the event of failure in the production system so that integration continues seamlessly when using the backup system and there is little to no effect on clinical workflow during this downtime. Fig. 5.3 shows a possible configuration of servers and routers for a development system *(left)*, production system *(center)*, and backup system *(right)*.

Both IT and clinical engineering resources are used to build and develop these systems. Work must be done to translate the data fields and values that are sent from the medical records to fields within the medical record. For example, the monitor may send a value for heart rate that is called *HR*. This field needs to be mapped into the medical record field that is named *Heart Rate*. Although it may seem like these mappings are quite simple, in complicated patients with several invasive pressure measurements including intracranial pressure, central line pressure, and peripheral pressure measurements, ensuring that the right pressure measurement goes to the right medical record field is imperative. Additionally, confirmation of the units of measure is required to ensure that the values are seen as expected.

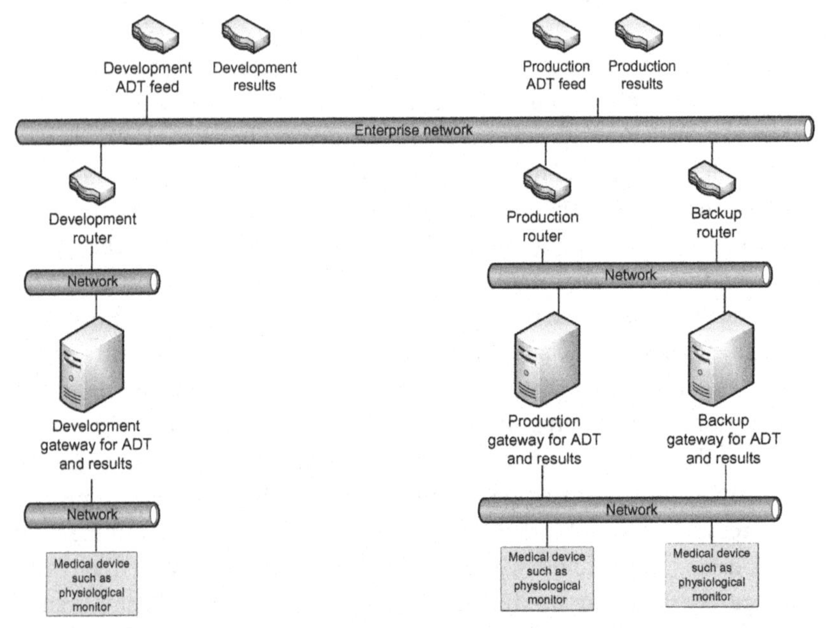

Figure 5.3 Possible network, server, and router configuration development system *(left)*, production system *(center)*, and backup system *(right)* to integrate medical devices such as physiologic monitors to patient medical records.

Push workflows

Push workflows usually occur when a test, such as ultrasonography or magnetic resonance imaging (MRI), is needed on a patient. The clinician or provider creates an order within the patient's medical record detailing the test that is required. Medical devices such as ultrasound scanners then use an ADT interface and software called a *worklist* to populate a list of the tests that are ordered. The technician performing the test then reviews the worklist, selects the order for the patient he or she is with, verifies the patient information (to ensure that this is the correct order), and performs the test. Because the test was done under an order, the test results are automatically paired with the results and made available in the patient's medical record.

The network configurations for this order-based workflow are similar to Fig. 5.3. This workflow is generally used when orders are needed to provide details about the test (e.g., test site, laterality). It also allows for billing to be easily integrated as once a test is completed, not only are the results tied to the patient medical record, but a bill for the test can be automatically generated.

Picture archiving and communication system integration

A picture archiving and communication system (PACS) is a technology that provides archiving of images from multiple modalities such as MRI, computed tomography, ultrasonography, and mammography. Similar to HL7 standards, PACS systems use a Digital Imaging and Communications in Medicine (DICOM) standard that consists of a file format and header information that identify the patient and the examination.

PACS systems were developed as part of the transition from hard-copy images such as film archives. These systems allow remote access to images, allow integration of images to medical records, and can act in concert with a radiology information system (RIS) to help manage the orders, worklists, and other workflows needed in a radiology department.

The workflow for this integration starts at the device that generates an image. Fig. 5.4 shows the configuration for this workflow. The DICOM-wrapped image is sent through the network to a gateway or workstation that validates the image and patient information. Once the quality check is completed, the image is sent to the PACS for storage. From the PACS servers, physicians can call up the images from their reading stations, or hyperlinks to images can be imbedded in the patient's electronic medical

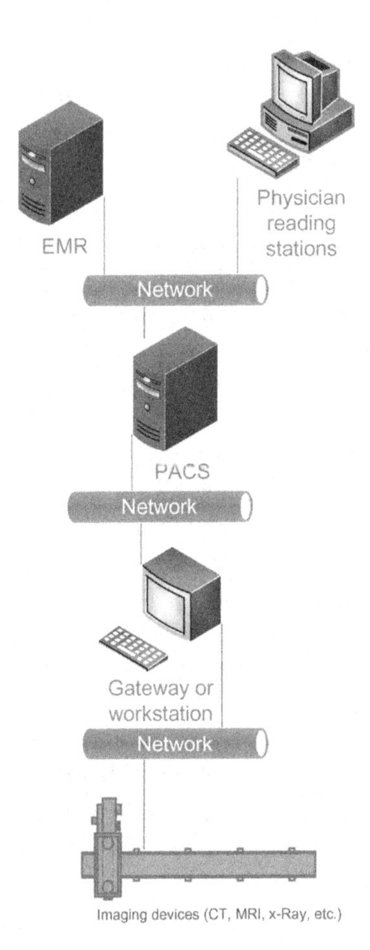

Figure 5.4 Possible PACS configuration to allow for radiology storage and archive of images.

record. Depending on where the network and how network availability is architected, clinicians can retrieve images remotely and have access to scans in near real time.

Most existing PACS systems focus on imaging devices as they use the DICOM standard. However, hospitals and healthcare systems generate other images from cameras, such as in a dermatologist's office, or from a rental camera, such as in an ophthalmologist's office. Because these images cannot be stored as image files directly in the medical record, hospitals are looking at ways to apply the DICOM standard to these images so that they can be stored in the same manner as the PACS images.

Some vendors that accept these types of multi-departmental images have renamed their PACS systems as a vendor-neutral archive (VNA) so that the images are not solely aligned with radiology departments. As long as the images submitted can be DICOM-compliant, any type of image can be stored in these VNAs and be accessible in the same manner that PACS images are available.

Laboratory information systems

Similar to how imaging devices are attached to an RIS and a PACS, laboratory analyzers can be integrated into a laboratory information system (LIS). These systems help align orders to laboratory samples and then take results provided by the laboratory analyzers and send them back to the central LIS. The LIS can then be integrated to the medical record so that results from the laboratory can populate into the patient's medical record. Again, the backbone for the LIS is the ADT system. The ADT system matches the patient with the results so the correct results will be transmitted to the patient's record. This ADT interface is critical in two separate places within the flow of samples.

First, when the laboratory sample is taken, the ADT interface is used to match the patient with the sample. This workflow usually requires a nurse or phlebotomist to scan the patient's identification wristband (within the hospital) or verify the patient (in an outpatient setting). This identification is transmitted to a barcode label that is placed on the sample. The sample is then sent to the laboratory, where it is processed. The results are matched with the barcode to ensure that the results and the patient are matched. The results are sent to the LIS, which then interfaces to the electronic medical record.

Decision support algorithms

With all of the data being generated and logged in medical records, companies and organizations have been using data-mining techniques to develop best practices for healthcare providers. These best practices have been implemented using tools such as clinical decision support (CDS) algorithms. CDS provides clinicians, staff, patients, and other individuals with knowledge and person-specific information, intelligently filtered or presented at appropriate times, to enhance health and healthcare (Office of the National Coordinator for Health Information Technology, 2018).

One example of a decision support is a sepsis alert. Sepsis is a condition that occurs when the body responds to an infection in a manner that injures its own tissues and organs. It can be life-threatening and lead to shock, multiple organ failure, and death, especially if it is not recognized and treated promptly. Several different sepsis alerts use discrete measures such as heart rate, non-invasive blood pressure, respiratory rate, glucose level, white blood cell count, temperature, oxygen needs, creatinine levels, and other metrics to identify if a patient is at risk for developing sepsis. These decision support systems use metrics entered in the medical record to assess whether patients are at risk. If a risk is identified, then an alert is triggered to the clinician.

Secondary device notification systems

Secondary device notification systems are systems in which alarms from primary devices such as monitors, ventilators, or other devices are sent to a secondary device such as an iPad, pager, or smartphone. These systems enhance workflow because clinicians such as physicians and nurses do not need to be near the device to receive notifications. These systems then use software called *middleware* to create rules to suppress, translate, escalate, and/or communicate these alarms to a secondary device (see Fig. 5.5). Development of these rules dramatically affects what and how alarms are sent (Jacques, 2017).

When implementing a secondary device notification system, the type of device that is selected to receive the messages from the middleware significantly affects the data the end user sees. Choice of a pager, smartphone, or tablet device affects the type of data (text vs waveforms), context of the alarms (single vital vs multiple vitals), size of the text, size of the waveform, and how the information is conveyed (Jacques, 2017). Pagers may limit the number of characters that can be delivered in one message and generally cannot support images such as waveforms.

Figure 5.5 Alarm flow for alarms produced at the primary device that flow through the middleware (software) to the secondary device for secondary alarm notification (Jacques, 2017).

For each device that sends data to the middleware system, an analysis of which alarms are generated should be passed through to the end user via the secondary device. All alarms generated by a primary device may not need to be sent to the clinician's secondary device as this may exacerbate alarm fatigue by clinical staff. Implementing a secondary device notification system will not reduce alarm fatigue unless proactive decisions are made to limit the number and type of alarms that are sent through the middleware (Jacques, 2017). Normally, hospitals will send only critical or life-threatening alarms through the middleware to the secondary device.

In addition to determining which alarms should be transmitted, teams implementing secondary device notification systems also need to determine to whom the alarms go and what the escalation process is if alarms are not acknowledged. Understanding which nurses and physicians are caring for each patient is imperative to ensure that the notifications go to the right person. Evaluation of workflows and actions clinicians must take when notified is also needed to ensure that the notifications aid in meeting clinician needs instead of hindering how clinicians take care of patients.

Telemedicine

Telemedicine is the use of telecommunications and IT to provide clinical healthcare remotely. The high-fidelity communication allows for transmission of medical, imaging, and health informatics data from one site to another. It has been used in many applications to overcome barriers such as distance and resources.

In rural settings, access to a wide variety of medical services can be a challenge because of resource limitations or lack of trained medical staff. Telemedicine options allow for access to specialty services in remote locations by connecting rural community health centers and hospitals to services that cannot be resourced in these locations.

Telemedicine can also be used in critical care and emergency situations. When hospitals are not equipped to provide the level of care needed, telemedicine services can be used to augment the available services. For example, tele-stroke programs aid hospitals in the identification and treatment of stroke victims, and tele-ICU programs remotely monitor patients in the intensive care unit to alert local caregivers to changes in the patient's status.

Other telemedicine programs can provide primary care to help alleviate unnecessary emergency department visits. Most hospitals struggle to get non-emergency patients to use primary care visits as a means of receiving their healthcare as opposed to going to an emergency department. Some hospital systems have implemented virtual visits that can be accessed via a smartphone app or a computer for patients who need to see a primary care physician for issues such as a cold or flu.

Medical devices at home

As discussed in Chapter 2, Healthcare technology basics, medical technology, both simple and complex, is utilized in the delivery of patient care in the home setting. The use and monitoring of devices is facilitated by the interconnectedness of homes and healthcare facilities through computer networks. The IoT has opened the opportunity to use more medical devices outside the hospital and healthcare settings. By connecting medical devices such as blood glucose monitors, heart rate monitors, and patient scales to the Internet, an entire new application of home healthcare has been created. More monitoring of healthcare conditions, such as diabetes and heart disease, can be done in a patient's home instead of requiring patients to travel to their healthcare provider for regular checks. This ability to monitor data over time also gives healthcare providers a better view into the overall health of the patient instead of relying on data from only a few visits per year.

This technology application is new and changing rapidly, however the benefits are expected to be increased innovation, improved collaboration with patients and caregivers, increased cost savings, and improved patient outcomes. With these benefits, the dangers of erroneous physiologic data storage or communication as well as the interruption of data communication are risks to balance. Unintended errors as well as nefarious actions must be evaluated and mitigated.

Cybersecurity, threats, vulnerabilities, and mitigation

Cybersecurity is an emerging component of the clinical engineering profession. Challenges include the evolution of threats occurring faster than policies can be adopted or training can be provided. In general, a lack of understanding about the types of cyberthreats and tools for detection and mitigation, coupled with a lack of resources dedicated to cybersecurity are tremendous challenges for the healthcare technology management (HTM) profession. ANSI/AAMI/IEC 80001-1:2010, *Application of risk management for IT networks incorporating medical devices*, is a standard that explores the activities that are required to minimize risk and promote system security. The standard focuses on the safety, effectiveness, and system security of healthcare information. The role of a medical IT network risk manager is identified and described.

Ransomware

Ransomware is a type of malicious software, called *malware*, that is designed to deny access to computer systems until a ransom is paid. Ransomware typically spreads through phishing emails or by visiting an infected website (US-CERT, 2019a). The US Computer Emergency Readiness Team (US-CERT), a part of the US Department of Homeland Security's CyberSecurity and Infrastructure Security Agency (CISA), recommends the following precautions to protect against ransomware:

- Update software and operating systems with the latest patches. Outdated applications and operating systems are the target of most attacks.
- Never click on links or open attachments in unsolicited emails.
- Back up data on a regular basis. Keep it on a separate device, and store it offline.
- Follow safe practices when browsing the Internet (US-CERT, 2019a).

In addition, CISA also recommends that organizations employ the following best practices:

- Restrict users' permissions to install and run software applications, and apply the principle of "least privilege" to all systems and services. Restricting these privileges may prevent malware from running or limit its capability to spread through a network.
- Use application whitelisting to allow only approved programs to run on a network.
- Enable strong spam filters to prevent phishing emails from reaching end users, and authenticate inbound email to prevent email spoofing.
- Scan all incoming and outgoing emails to detect threats and filter executable files from reaching end users.
- Configure firewalls to block access to known malicious IP addresses (US-CERT, 2019a).

Denial-of-service attack

A denial-of-service (DoS) attack is meant to shut down a machine or network, making it inaccessible to its intended users. DoS attacks accomplish this by flooding the target with traffic or by sending the target information that triggers a crash. In both instances, the DoS attack deprives legitimate users (i.e., employees, patients, physicians) of the service or resource they expected (US-CERT, 2019b).

Though DoS attacks do not typically result in the theft or loss of significant information, they can cost the hospital or healthcare system a great deal of time and money to handle. There are two general methods of DoS attacks: flooding services and crashing services. Flood attacks occur when the system receives too much traffic for the server to buffer, causing it to slow down and eventually stop (US-CERT, 2019b). Other DoS attacks simply exploit vulnerabilities that cause the target system or service to crash. In these attacks, input is sent that takes advantage of bugs in the target, which subsequently crash or severely de-stabilize the system so that it cannot be accessed or used (Palo Alto Networks, 2019).

An additional type of DoS attack is the distributed denial-of-service (DDoS) attack. A DDoS attack occurs when multiple systems attack a single target in a synchronized way. The essential difference is that instead of being attacked from one location, the target is attacked from many locations at once (US-CERT, 2019b). The distribution of hosts that define a DDoS provides the attacker with multiple advantages:

- The attacker can leverage a greater volume of machine to execute a seriously disruptive attack.
- The location of the attack is difficult to detect because of the random distribution of attacking systems (often worldwide).
- It is more difficult to shut down multiple machines than one.
- The true attacking party is very difficult to identify, as they are disguised behind many (mostly compromised) systems (Palo Alto Networks, 2019).

Modern security technologies have developed mechanisms to defend against most forms of DoS attacks, but because of the unique characteristics of DDoS, it is still regarded as an elevated threat and is of higher concern to organizations that fear being targeted by such an attack.

Formjacking

Data interception or manipulation associated with online form entry is a multi-disciplinary problem that can affect healthcare institutions. The process is relatively simple from a programming point of view, and common attacks occur on popular sites such as ticket vendors and airlines. Simple security measures can prevent this type of attack but do require proactive prevention. Staff training and vigilance can help mitigate this type of attack.

Cybersecurity of medical devices

IT security issues associated with medical devices are of extreme importance for healthcare institutions. To be clear, these risks do not appear in isolation. Medical device cyber-risks are tightly coupled with institutional challenges and IT hazards. However, although IT professionals have experience and best practices associated with network threats and mitigation, the issues associated with devices and their functionality are slowly emerging and communicated equally as slowly. One of the first medical device vulnerabilities was identified in 2011 when Barnaby Jack demonstrated hacking of a friend's insulin pump at a security conference. Although gimmicky events have surfaced and other potential device vulnerabilities have been identified, including pacemakers, brain stimulators, and infusion pumps, at the time of publication no medical device nefarious activities have been identified. However, no hospital wants to be the site of a hacking incident that results in patient harm.

Commonly connected devices include electroencephalogram and ultrasound machines, ventilators, pacemakers, dialysis units, and infusion pumps. Many devices that utilize software lack basic security techniques like password controls in an effort to balance patient safety with cybersecurity. Emergency treatment cannot rely on limited knowledge of a password. Stealing credentials as a threat is relatively simple if the credentials are non-existent or readily available.

In an effort to mitigate challenges in the future, medical device manufacturers are asked to design technology with cybersecurity as part of the planning process. A 2005 guidance document by the US Food and Drug Administration (FDA) focused on cybersecurity for networked medical devices containing off-the-shelf software (FDA, 2005). In 2018, the FDA issued a guidance document associated with the design of medical devices to include the management of cybersecurity. However, these guidelines are intended for manufacturers designing the next generation of devices and do not mitigate the vulnerabilities associated with current technology. Most manufacturers see it as too difficult or expensive to secure the existing and legacy devices and leave most risk mitigation to the hospital's IT departments.

Guidance to detect and mitigate risk associated with medical devices for HTM professionals is available. In 2016, the Association for the Advancement of Medical Instrumentation (AAMI) published AAMI TIR57:2016, *Principles for medical device security—Risk management*. The Joint Commission has incorporated cybersecurity awareness into several Environment of Care elements of performance. Accreditors recommend that clinical engineers integrate cybersecurity activities as part of equipment life-cycle planning and within the preventative maintenance procedures associated with a particular device.

Cybersecurity risk analysis and prevention should be incorporated as a key component of all stages of the life cycle of devices. First, the identification of risks associated with a particular device is critical, including physical weaknesses, software holes, or Internet connection challenges. Checklists and other tools such as the Medical Device Risk Assessment Platform (MDRAP) and Zingbox IoT Guardian can assist clinical engineers in device risk management. Vulnerability assessment can be shared knowledge across institutions for some device categories. During this assessment, the evaluation of software currency or other preventative measures is useful and ensures compliance with medical device manufacturer recommendations. Second, careful tracking of device-specific information, such as the IP address and operating system, should be recorded in the computerized maintenance management system for review during overall institutional risk assessment. Third, clinical engineers should be sure to educate medical device users to understand the need for updates or vigilance and to practice techniques to reduce risk. Lastly, disposal of devices should be carefully managed to ensure that patient data are removed.

In 2017, the AAMI Wireless Strategy Task Force offered a list of 10 ways to mitigate risk for cyberattacks on medical devices (AAMI Wireless Strategy Task Force, 2017). These include many common recommendations associated with devices that connect to the Internet, such as using strong passwords, encrypting data, and performing periodic reviews of vulnerabilities. Fundamentally, many medical devices are equivalent, from a security perspective, to other technologic tools. However, the risks associated with adverse events can be significant or critical.

Collaboration between IT and HTM, long a thorny partnership, is critically needed in the cybersecurity battle. Ensuring that software updates are timely, thorough, and well-documented is crucial in ensuring patient safety. In addition, planning for a cyberevent can create an incident response intervention that may mitigate a negative impact. Increasing the awareness of vulnerabilities and threats associated with medical devices through clinical departments and the administration can be a challenge until an adverse event associated with a medical device occurs.

Cybersecurity incident response plans

Risk identification and remediation can lead to proper integration of medical devices to networks that helps mitigate the risks brought on by cyberattacks. However, prevention is not the only planning that needs to be done between IT and clinical engineering departments. Cybersecurity events occur every day, and most hospitals have well-defined incident response plans that help departments navigate through incidents and respond to threats in the moment. An incident response plan should detail resources that are available to respond to an incident as well as their roles and responsibilities. These details help outline the team and the chain of command for decisions and responses. The plan should also categorize incidents. Most cybersecurity departments use definitions by the National Institute of Standards and Technology (NIST), a division of the US Department of Commerce, as this helps them align with external partners. In addition to categorization, prioritization should also be defined. This helps resources align priorities during critical times. Finally, the plan should detail the process for handling incidents including notifications, triage, containment, eradication, recovery, and post-incident activities.

Once this plan is developed, generally it is tested using real-life scenarios during a tabletop exercise. These tabletop exercises help determine the feasibility and usefulness of the plan. Once the exercise is concluded, lessons learned can be molded into an after-action review. After-action reviews of tabletop exercises can then help refine and adapt the plan. This continuous improvement activity helps ensure that plans are actionable. After-action reviews can also be developed using actual events and again are another opportunity to enhance and refine incident response plans.

Abbreviations

ADT system admission, discharge, and transfer system
CISA CyberSecurity and Infrastructure Agency
DoS attack denial-of-service attack
DDoS distributed denial-of-service attack
HL7 standards Health Level 7 International standards
ITIL Information Technology Infrastructure Library
IoT Internet of Things
LIS laboratory information systems
OSI model Open Systems Interconnection model
PACS picture archiving and communication system
RIS radiology information system

References

AAMI Wireless Strategy Task Force. (2017). Cyber vigilance: Keeping healthcare technology safe and secure in a connected world. *Biomedical Instrumentation and Technology, 51* (s6), 10–16.

Cannon, D. (2011). ITIL Service Strategy. ITIL lifecycle suite, 2011 edition (2nd ed.). The Stationery Office.

Hunnebeck, L. (2011). ITIL Service Design. ITIL lifecycle suite, 2011 edition (2nd ed.). The Stationery Office.

Jacques, S. (2017). Factors that affect design of secondary alarm notification systems. *Biomedical Instrumentation and Technology, 51*(s2), 16–20.

Lloyd, V. (2011). ITIL Continual service improvement. ITIL lifecycle suite, 2011 edition (2nd ed.). The Stationery Office.

Office of the National Coordinator for Health Information Technology. (April 10, 2018). *Clinical decision support.* Retrieved from <https://www.healthit.gov/topic/safety/clinical-decision-support>.

McGonigle, D. & Mastrian, K.G. (2012). Nursing informatics and the foundation of knowledge (2nd ed.). Burlington, MA: Jones & Bartlett Learning.

Microsoft. (April 20, 2017). *Windows network architecture and the OSI model.* Retrieved from <https://docs.microsoft.com/en-us/windows-hardware/drivers/network/windows-network-architecture-and-the-osi-model>.

Palo Alto Networks. (2019). *What is a denial of service attack (DoS)?* Retrieved from <https://www.paloaltonetworks.com/cyberpedia/what-is-a-denial-of-service-attack-dos>.

Rance, S. (2011). *ITIL service transition. ITIL lifecycle suite, 2011 edition* (2nd ed.). The Stationery Office.

Steinberg, R.A. (2011). *ITIL service operation. ITIL lifecycle suite, 2011 edition* (2nd ed.). The Stationery Office.

US Computer Emergency Readiness Team. (2019a). *Ransomware.* Retrieved from <https://www.us-cert.gov/Ransomware>.

US Computer Emergency Readiness Team. (2019b). *Understanding denial-of service attacks.* Retrieved from <https://www.us-cert.gov/ncas/tips/ST04-015>.

US Food and Drug Administration. (2005). *Cybersecurity for networked medical devices containing off-the-shelf (OTS) software.* Retrieved from <https://www.fda.gov/regulatory-information/search-fda-guidance-documents/cybersecurity-networked-medical-devices-containing-shelf-ots-software>.

CHAPTER 6

Facilities management

Introduction

Without proper management of a facility, inadvertent exposures to environmental opportunistic pathogens may result in infections with significant morbidity and/or mortality. Pathogens such as *Aspergillus* and *Legionella* can grow and spread in poorly maintained water systems, and airborne pathogens such as *Mycobacterium tuberculosis* and varicella zoster virus can spread due to lack of management of HVAC systems. Lack of adherence to established standards and guidance, including water quality in dialysis, proper ventilation for operating rooms, and proper use of disinfectants for cleaning, can result in adverse patient outcomes in healthcare facilities (CDC, 2017b).

There are many standards, regulations, and codes that affect hospitals. Many states and local health departments have their own regulations, however, the Joint Commission, the Centers for Medicare and Medicaid Services (CMS), and the National Fire Protection Association (NFPA) have other regulations and standards used in design and construction of facilities. The most frequency used codes include:

- NFPA 70: *National electrical code;*
- NFPA 99: *Health care facilities code;*
- NFPA 110: *Standard for emergency power standby systems;* and
- NFPA 111: *Standard on stored electrical energy emergency and standby power systems.*

These codes specify a number of requirements including ground fault interrupters in wet locations, the automatic transfer of power between switches, and the medical gas requirements. There are also different types of generators, each with its own fueling requirements and procedures. Regulations also discuss how and when to preventively maintain and test primary and backup systems.

Electrical power

A hospital's electrical system is required to provide reliable, interruption-free power constantly. The complexity of diagnostic and treatment

Introduction to Clinical Engineering
DOI: https://doi.org/10.1016/B978-0-12-818103-4.00006-5

equipment utilized in a hospital make it highly reliant on a secure and consistent power supply.

Most hospitals rely on the electrical grid for its primary power, but they cannot rely solely on this power as any event, including weather-related incidents, brownouts, or failure in the electrical systems, may interrupt this power. Therefore, hospitals must also have backup systems that may include generators to continue the power in case of a primary power outage. However, backup systems are only designed to continue power for the most critical areas and equipment; they are not meant to keep the entire hospital powered. Hospitals use different designations for power outlets to indicate whether they are supported by backup or emergency power. White outlets are designated as primary outlets. Generally red outlets are designated as emergency outlets (some hospitals may use orange outlets because red outlets are used for fire alarm systems). Any equipment plugged into white outlets when the transition from main to backup power is made will not continue to have power; only the equipment plugged into red outlets will continue to function on emergency power.

The emergency power system only consists of circuits essential to life safety and critical patient care. Each branch of the emergency power system is served by one or more transfer switches so that the transferring does not overload the generation. The design and implementation of the electrical system must account for the loads generated and transferred during the power outage and the time it takes to switch from normal to backup power. Depending on the backup system installed, this can be anywhere from a few seconds to a few minutes. Hospital equipment that is very sensitive to power interruptions may also be placed on an uninterruptable power supply to even out this transition and prevent interference with the function of the device.

The emergency system is made up of two mandatory branches: the life safety branch and the critical branch. The life safety branch provides the power supply for lighting, receptacles, and equipment including elevators, exit signs, and automatic doors. The critical branch is used for fixed equipment and special power circuits such as telephone equipment closets, nurse call systems, blood/tissue banks, red outlets in patient or treatment rooms, and other isolated power systems (NFPA, 2015). Additionally, life safety and critical branches are wired separately from all other hospital wiring. They are placed in different raceways, boxes, and cabinets so that the number of receptacles can be limited, minimizing the risk for a branch circuit outage.

Power and backup power systems must be maintained regularly. The emergency power supply system and its components must be inspected at least weekly, and generator testing and transfer switch tests must be done under load at least once every 30 days. Additionally, a 4-hour generator test must be done under load once every 36 months (NFPA, 2019). Records of all tests must be kept and are reviewed by regulatory agencies during inspections.

Plumbing systems and water quality management

Hospital plumbing systems are designed and built to meet the various needs of the organization. The equipment needed has various requirements and different maintenance considerations than normal plumbing installations. Most office buildings or hotels deliver hot water at 120°F to 140°F, however, in hospitals, the water temperature for standard showering and hand washing is limited to 110°F. Additionally, the water temperature on psychiatric floors and in areas with dementia patients is limited to 100°F. Some examples of other special departmental requirements are described in the following sections.

Dialysis systems

Dialysis systems have very rigorous regulations about water quality. The US Environmental Protection Agency (EPA) sets standards for drinking water quality under the Safe Drinking Water Act (SDWA) (EPA, 2017), however water used for dialysis must be ultrapure. The standard for ultrapure water, along with the standards for dialysate quality and equipment, are set by the Association for the Advancement of Medical Instrumentation (AAMI) in ANSI/AAMI/ISO RD-52:2004, *Dialysate for hemodialysis* (AAMI, 2004). There are two options for meeting these regulations: use of a portable water filtration cart or installation and use of a reverse osmosis (RO) system.

An RO system uses a pump to push water through a semipermeable membrane that removes contaminants. The system may also have several filters, such as a sediment filter that traps large particles and a carbon filter that absorbs chemicals. Finally, an RO system may have a softener that removes calcium and magnesium from the water. Each system is designed and built based on the feed water for the system and the steps needed to ensure that the resulting water is ultrapure. The resulting ultrapure water is then mixed with the dialysate for the dialysis treatment. Waste or reject water that contains bacteria and particles from the filtration is discarded, usually down a drain.

Because these systems tend to be large and require significant up-front capital costs, maintenance, and monitoring, they are generally installed only in larger clinics dedicated to dialysis treatments.

Portable water filtration systems are typically smaller RO systems that can be hooked up to municipal tap water. Because of their size, they can be used in a patient's home, a clinic, or a hospital.

Both RO systems and portable systems require regular maintenance and water testing to ensure that the filters and membranes are functioning properly and producing the required ultrapure water.

Sterile processing systems

Within a hospital, this department is responsible for the cleaning and sterilization of surgical tools and instruments. Because of the large volume of water (and steam) needed and discarded, designs to accommodate the load must be incorporated into the system.

Additionally, specialized needs for sterilizer systems may include requirements for water at a certain temperature (usually between 160°F and 180°F). To obtain sterilization, it is critical that the minimum temperature requirement be met throughout the sterilization cycle. Certain sterilizers may also require backflow prevention, temperature-mixing valves, and filters. These sterilizers may be placed in sterile processing departments, operating rooms, and other procedural areas. Regular preventive maintenance of the water system will aid in the uptime of the sterilizer units.

Steam generation for sterile processing is generally done using the domestic hot water system, control valves, and a condensation pump. This system requires preventive maintenance to ensure that calcium deposits do not reduce the effectiveness of the system. Additionally, if gas generators are used to create the steam, then a storage tank will be required to preserve the gas on-site.

Laboratory systems

Hospital laboratories require different types of purified water to complete pathology and other tests. Depending on the types of analyzers and tests performed by the laboratory, RO water and deionized water may be required.

Additionally, the design of the plumbing system may require considerations for the waste side of the plumbing system. Certain laboratory systems may require installation of backflow preventers so that there is no contact with the water after it is disposed of down the drains. Additionally, point-of-use acid-neutralizing systems, antifoam systems, and/or drain covers may be required to deal with wastewater and chemicals being disposed of.

Emergency plumbing systems

Throughout the hospital and within outpatient clinic locations, emergency plumbing systems such as eyewash stations and emergency showers may be required based on the work done in that location.

The Occupational Safety and Health Administration (OSHA) requirements for emergency eyewashes and showers, found at 29 CFR 1910.151 (c), specify that "where the eyes or body of any person may be exposed to injurious corrosive materials, suitable facilities for quick drenching or flushing of the eyes and body shall be provided within the work area for immediate emergency use." An eyewash and/or safety shower is required where an employee's eyes or body could be exposed to injurious corrosive materials as indicated by the safety data sheet (OSHA, 2019).

As the OSHA regulations are vague in what defines "suitable facilities," the American National Standards Institute (ANSI) provided more guidance by establishing a standard that covers eyewash and shower equipment—ANSI Z358.1:2014, *American national standard for emergency eyewash and shower equipment*. This standard includes strict ANSI regulations on temperature, flow rates, spray patterns, and types of devices (e.g., hands-free) that must be followed (ANSI, 2014). In general, emergency equipment must be within 10 seconds (or approximately 55 feet) from the location of the hazard. The path must be as straight as possible and free from obstructions. The temperature of the delivered water should be tepid (between 60°F and 100°F). Water supply for shower devices must be sufficient to supply at least 20 gallons per minute for 15 minutes. Eye and face washes must provide low-velocity flow that completely rinses eyes and face but is not injurious to the user (ANSI, 2014).

Water quality management

In addition to managing the incoming and outgoing water needs of the institution, facilities departments are also responsible for maintaining the quality of the water within the hospital. Water quality management is necessary to reduce the risk of *Legionella* in areas where patients are exposed to water droplets resulting from medical procedures (e.g., surgery, hydrotherapy) and for patient populations that are vulnerable to infection (e.g., bone marrow transplant patients, oncology patients).

In general, the principles of effective water management include:

- maintaining water temperatures outside the ideal range for *Legionella* growth;
- preventing water stagnation;
- ensuring adequate disinfection; and
- maintaining premise plumbing, equipment, and fixtures to prevent scale, corrosion, and biofilm growth, all of which provide a habitat and nutrients for *Legionella* (CDC, 2019b).

The Centers for Disease Control and Prevention (CDC) has developed a guide to implementing industry standards. The guide recommends creating a continuous water management program that:

- establishes a team;
- describes the building water system;
- identifies where *Legionella* could grow and spread;
- decides where control measures should be applied and how to monitor them;
- establishes ways to intervene when control measures are not met;
- ensures that the program is running as designed and is effective;
- documents and communicates all activities (CDC, 2017a).

The guide allows flexibility for healthcare organizations of all sizes and provides a continuous process approach that helps reduce the risk for growing and spreading *Legionella*.

Medical gas systems

Medical gas systems have several layers of regulation and oversight from the CMS and OSHA as well as the NFPA 99 code. The medical gas system generally has a bulk container, manifolds, compressors, and pumps that attach to the piping system within the hospital. The tanks of medical gases are either within a special room in the hospital, or more often located outside the building (DiMarco, 2017).

Oxygen

Liquid oxygen boils at normal temperatures, and the gaseous phase is piped at a regular pressure into the hospital building, where it is measured and regulated. An oxygen system requires a main system and backup system with a shutoff valve and pressure monitors at the entry point to the hospital. Alarms are set to ensure that gas is at the right pressure and provide a measurement of the volume of the supply. These alarms are generally linked into the automated building system, which notifies facilities and security if there is an issue. Should the main system fail, a hospital is required to have at least a one-day supply of oxygen on hand as well as a one-day backup supply (DiMarco, 2017).

Once the oxygen is in the piped gas system, it flows through a system of special copper pipes that deliver the pure oxygen to the walls and outlets within the hospital. The system design includes isolated sections of piping that allow it to be maintained and repaired without taking down the entire system. Zone valves control the oxygen flow to these isolated sections—usually a set of patient rooms or a floor. These zone valves have visual gages to identify the pressure in the system and are usually set into the wall so that only authorized personnel can access them. The oxygen flows from the zone valves to the outlets in the patient areas for use.

The wall outlets have specific quick-connect valve fittings that come in seven different configurations. The fittings and flexible hoses that connect are colored green, indicating that the connection is for oxygen. When the patient's cannula or oxygen mask is connected to the wall oxygen, generally a volume/pressure regulator is attached to the wall fitting to control the amount of oxygen being produced.

Vacuum

Vacuum pumps create a vacuum by removing the gases from the receiver tank and forcing the compressed air outside, leaving a vacuum in the receiving tank that is connected to the house piping. System designs are required to be redundant because the loads vary as the system is used. Various controls maintain the pressure and alarm to the master building system if the pressure is not adequate.

Vacuum connections in patient care areas are connected by outlets with the same quick-connect valve fittings as oxygen. However, each gas fitting has its own unique hose connection. This is done to prevent cross-connection between medical gases and to increase patient safety. Vacuum outlets are color-coded white. As with oxygen, regulators are plugged into the room outlet to manage the strength of the vacuum and to act as the on/off valve within the room.

In operating rooms and other treatment locations, vacuum systems are used to draw fluids and solids out of the body, however this results in an increased likelihood of accumulating these fluids/solids within the piping and possibly in the receiving tank and pump itself. If this occurs, the lines, inlets, and even the plumbing can occlude and reduce the pressure that can be delivered to the room. When this occurs, the system must either be cleaned or the piping replaced to remove the fluids/solids. To mitigate fluids entering the vacuum system, gravity traps and canisters are attached to the vacuum inlet to collect the fluids. These systems reduce the likelihood of intrusion into the vacuum system but cannot completely eliminate the infiltration because some particles can be aerosolized (DiMarco, 2017).

Waste anesthetic gas disposal

A waste anesthetic gas disposal (WAGD) system is a special vacuum system used to reduce the risk for too much anesthetic gas building up in surgery and procedure areas. During procedures, anesthesiologists mix anesthetic gases to sedate the patient. Not all anesthetic gases are consumed by the patient, and these gases can return to the mask when the patient exhales (DiMarco, 2017). Exposure of even minimal amounts of anesthetic gases over time can be harmful to the staff performing the surgery, so the WAGD system draws the exhaled gas away from the patient and staff. These outlets are color-coded purple, and fittings are again different from the fittings for other gases to prevent cross-connections.

A WAGD system can be designed so that vacuum pressure is maintained separately from the vacuum system within the operating room, or the system can be merged with the vacuum system outside the operating room. The volume of surgeries performed will inform whether the systems need to be totally separate or can use the same plumbing. The focus of the system is to draw the gases away from the patient and pump them outside the building, usually via a roof vent. This roof vent must be in a different location from other air intakes so that the gases are not pumped back into the hospital. The pressure within the system must be significant enough to pull the anesthesia/oxygen mix through the system and out rather quickly as the mix of anesthesia and oxygen is flammable.

Medical air

Medical air is created in a central location to provide air to patients with the right humidity and composition for breathing, therefore the gas must meet US Pharmacopeia (USP) requirements. Generally, medical air is drawn from outside and compressed via a compressor and sent to a receiver tank. On its way to the receiver tank, the air is dried to the appropriate dew point and checked for carbon monoxide. When the air leaves the receiver tank as it travels through the gas system to the patient rooms, it is further filtered to ensure that all particles are removed (DiMarco, 2017). Medical air fittings are color-coded yellow and again have different hose connections to prevent cross-connections.

Medical air is another system that must be available at all times, so the compressor system and air dryer must be redundant. Pumps and dryers must be inspected regularly to ensure that they are functioning properly and removing water from the compressed air.

Heating, ventilation, and air-conditioning systems

Mechanical systems including HVAC (heating, ventilation, and air-conditioning) systems in hospitals are designed and maintained to a higher standard than those in typical buildings. Increased system design demands including infection prevention, patient comfort, and regulations cause these systems to be quite different from normal systems. These systems are regulated by standards established by a number of organizations including, but not limited to, the American Society for Health Care Engineering (ASHE), the National Fire Protection Association (NFPA), ANSI, and the American Society of Heating, Refrigerating and Air-Conditioning Engineers (ASHRAE).

HVAC systems in healthcare facilities are designed to:

- maintain the indoor air temperature and humidity at comfortable levels for staff, patients, and visitors;
- control odors;
- remove contaminated air;
- facilitate air-handling requirements to protect susceptible staff and patients from airborne healthcare-associated pathogens; and
- minimize the risk for transmission of airborne pathogens from infected patients (CDC, 2017b).

The general method used to remove particulates from the air is filtration. Several filter banks may be used to remove particulates including high-efficiency particulate air (HEPA) filters that are 99.97% efficient for removing particles greater than or equal to 0.3 μm in diameter.

As a supplemental air-cleaning measure, ultraviolet germicidal irradiation (UVGI) is effective in reducing the transmission of airborne bacteria and viruses as well as limiting the growth of vegetative bacteria and fungi. Although this method is effective for these types of infections, UVGI has a minimal impact on inactivating fungal spores (CDC, 2017b). Most commercially available UV lamps used for germicidal purposes are low-pressure mercury vapor lamps that emit radiant energy predominantly at a wavelength of 253.7 nm (CDC, 2017b).

Positive- and negative-pressure rooms

Positive and negative pressures refer to a pressure differential between two adjacent airspaces, (e.g., the room and the hallway). Air flows away from areas or rooms with positive pressure, while air flows toward areas with negative pressure.

Some rooms, such as isolation rooms, are set at negative pressure to prevent airborne microorganisms in the room from entering hallways and corridors. Other rooms housing severely neutropenic patients are set at positive pressure to keep airborne pathogens in adjacent spaces or corridors from coming into and contaminating the airspace occupied by such high-risk patients. Self-closing doors are mandatory for both of these areas to help maintain the correct pressure differential (CDC, 2017b).

According to the Facility Guidelines Institute/ASHRAE Standard 170-2013: *Ventilation of health care facilities*, rooms that should be negatively pressured include:

- emergency waiting rooms;
- triage rooms;
- radiology waiting rooms;
- airborne infection isolation rooms;
- pathology rooms such as cytology, histology, microbiology, and sterilizing laboratories;
- autopsy rooms;
- decontamination rooms for central medical and surgical supplies;
- soiled workrooms or soiled holding rooms;
- soiled linen and trash rooms;
- janitors' closets; and
- restrooms (Barrick & Holdaway, 2014).

Rooms that should be positively pressured include:

- operating rooms;

- delivery rooms;
- trauma rooms;
- newborn intensive care units;
- pharmacy;
- clean workrooms for central medical and surgical supplies;
- sterile storage rooms for central medical and surgical supplies (Barrick & Holdaway, 2014).

Root causes of issues with room pressurization may include an imbalance between supply and exhaust rates for the room. Supply and exhaust fans may not operate properly, or supply diffusers and return grilles within the room may be blocked. Other root causes may be imbalances in the system caused by recent renovations or changes in adjacent rooms (Barrick & Holdaway, 2014).

Air-conditioning for the operating room

ASHE Standard 170:2013, *Ventilation of Health Care Facilities*, requires that air in an operating room be introduced at the ceiling and exhausted at low wall grilles. A minimum of two low grilles, 8 inches above the floor, must be provided for return air. Airflow should be designed in a unidirectional or laminar air pattern to help reduce infection control issues (ASHE, n.d.). Laminar flow diffusers are generally installed to minimize other air patterns or mixing air currents in the room. ASHE standards indicate the area the diffuser should extend into and the percentage of area the primary diffuser supplies to nonlaminar areas (e.g., lights and gas columns).

Research by Kenneth Goddard in the 1960s started the interest in how the number of air exchanges in an operating room affects post-operative infection rates. This early research led to the current standard of 20 air exchanges per hour in an operating room. Higher exchanges of 40 air exchanges per hour have been historically used in orthopedic and open-heart operating rooms, although there is no conclusive evidence that this increased air exchange reduces surgical site contamination.

Building and renovation project management

Clinical engineering departments often partner with user departments during building renovations or construction. Clinical engineering's role is generally to aid in planning all equipment needs for the renovation/construction. In small projects, this work of defining equipment makes/models, quoting, ordering, and installing the equipment may lie primarily with the clinical engineering staff. In larger projects, architects may bring in third-party equipment planners to perform this function. These equipment planners may also provide CAD drawings of the placement of equipment in rooms, and they can aid in quoting, ordering, coordinating with vendors for their drawings, and installation planning. If organizations have equipment standards for the make/model of equipment they order, this process generally goes quite easily. If no standards exist, then the clinical engineering department or equipment planners usually work with the end users to identify needed requirements for their devices.

The American Institute of Architects (AIA) defines five phases of architectural services. Clinical engineering departments and equipment planners have different tasks during each stage.

In the *schematic design* phase, the architect works with end users to outline a program. This program describes the total scope of the work including details about the size, location, usage, and adjacency of rooms. The architectural teams may produce rough sketches and collect initial information about medical or other equipment that is architecturally significant. Architecturally significant equipment includes any device or system that requires water, power, data, or HVAC or that significantly affects the architecture or design (e.g., device weight, magnetic shielding). This early equipment planning for architecturally significant items may include vendors who need to provide specific drawings on their systems. Examples of such systems include magnetic resonance imaging or computed tomography machines, pneumatic tube systems, and device booms. At this stage, architects also meet with local code agencies to understand any zoning or code issues that need to be addressed and to obtain required permits.

In the *design development* phase, the architect further develops the designs including the civil, architectural, mechanical, electrical, and plumbing components.

Depending on the complexity of the project, partial drawings at the 50% complete or 75% complete stage might be created. These partial drawings are used to meet with end users and the mechanical, electrical, and plumbing (MEP) trades to coordinate pipe runs, power drops, and other workflow needs of the end users. During this phase, final equipment lists are created including the specific make/model of the equipment so that drawings can be produced with the equipment in place. This step ensures that required power, data, and HVAC systems are properly located in the same place the equipment is placed.

The *construction documents* phase includes development of the final plans with sufficient detail so that the construction manager and general contractor can build the project. These are also known as the *100% complete documents*. The architectural team uses these documents to achieve the final permits and prepare for construction. During this phase, equipment planners generally work with vendors to procure quotations and generate purchase orders for the items on the equipment list. Bidding for any equipment is also done during this time. Delivery dates and lead times provided during procurement generally align with the construction schedules so that the equipment is available at the right time during the construction phase.

In the *bidding and negotiation* phase, the architect works with the end user to bid on the project and selects a general contractor and a construction manager.

In the *construction administration* phase, the architect and owner review the job progress during the course of construction and review shop drawings to ensure that they conform to the design intent. Any additional coordination needed between the various trades is done, and any changes needed to the drawings are reviewed. Equipment is delivered, installed, and commissioned. Before use, regulatory bodies must review and approve the constructed spaces (AIA, 2009).

Emergency preparedness

Hospitals must be prepared for all types of emergencies. Departments within the hospital prepare for and run drills for a wide variety of events that may include but are not limited to:

- weather-related events;
- security events;
- large-scale events; and
- cyberattacks such as denial-of-service (DoS) attacks or ransomware attacks.

Weather-related events

Natural disasters and weather-related events include, but are not limited to, earthquakes, landslides, wildfires, floods, tornados, hurricanes, winter weather, extreme heat, and tsunamis. Location predisposes each hospital to some of these natural disasters, so hospitals must be prepared for each one. Unfortunately, the same plan does not necessarily work for each event. Some events can be predicted in advance, such as extreme heat or winter weather, and therefore preparations for appropriate staffing and supplies can be made ahead of time. Other types of events such as tornados and earthquakes occur with little warning and pre-planning. This is why emergency preparedness departments create separate plans for each type of weather-related disaster. These plans take into account the required personnel (both staff and leadership), supplies needed, and logistics to get these resources to the appropriate place.

Security events

Events that threaten the security of patients and staff may include a wide range of situations including bomb threats or active shooters to de-escalation of difficult patients. Hospitals and healthcare institutions have been seeing an uptick in workplace violence. From 2002 to 2013, incidents of serious workplace violence requiring days off for the worker to recuperate were four times more common in the healthcare industry than in the private industry (OSHA, 2015). Most of these events included physical assault/violence and abuse by patients against staff. In 2013, the Bureau of Labor Statistics reported that the highest rate of violent injuries was against psychiatric aids. Other high-risk areas include emergency departments, geriatrics, and behavioral health (OSHA, 2015). Hospitals have started to respond to this uptick using techniques such as staff training in de-escalation techniques and deployment of response teams that aid frontline staff in resolving issues.

More serious security events such as bomb threats and active shooters generally involve the security staff at the hospital as well as local and state police agencies. Response plans and training for staff on evacuation and shelter-in-place protocols are required. Local and federal agencies help train local staff and leaders in proper techniques and responses and are valuable resources for smaller hospitals.

Large-scale events

The CDC and other government agencies aid hospitals and health systems in preparation and planning for large-scale events such as bioterrorism emergencies, chemical or radiation emergencies, mass casualty events (e.g., plane crashes), and pandemics (e.g., flu). Similar to natural disasters, the location and size of each hospital or health system play a role in determining the risk of having to respond to each type of large-scale event. The Joint Commission and the CMS require annual drills that include not only single hospital responses, but coordinated responses from other entities as well. These drills help communities plan and practice their responses to these larger, more complex events.

Cyberattacks

Emergency planning around potential information technology (IT) attacks is critical to ensuring uninterrupted patient care. This emerging area of preparedness requires hospitals to explore and anticipate cyberattacks including ransomware attacks and DoS attacks. These attacks are specifically related to patient data rather than medical devices. As described in Chapter 5, Information technology, ransomware is a type of malicious software, called *malware*, that is designed to deny access to computer systems until a ransom is paid. The US Computer Emergency Readiness Team (US-CERT) recommends precautions to protect against ransomware. Ransomware typically spreads through phishing e-mails or by visiting an infected website (US-CERT, 2019a). A DoS attack is meant to shut down a machine or network, making it inaccessible to its intended users (US-CERT, 2019b).

Typically, IT departments are responsible for taking the needed precautions to prevent these attacks. However, with the increase in networked medical devices, all clinical engineers must be aware of these recommendations and actions as some of these roles and responsibilities will fall on the clinical engineering department. Additionally, good security procedures including safe Internet browsing, password changes, and vigilance for phishing attacks are the responsibility of all hospital personnel.

Although hospitals have experienced attacks associated with computer systems containing patient data, at the time of publication the US Food and Drug Administration (FDA) has not received a notification of a patient injury associated with a cybersecurity attack on a medical device. Certainly, patient care has been affected by the ransomware and DoS IT system incursions, but the changes were not directly associated with medical devices. Additional information about planning for medical device cyberthreats is provided in Chapter 5, Information technology.

Summary

Depending on the type of emergency, a structured command center may be stood up to be appropriately staffed to provide resources, manage communications, and continue to serve patients as needed. Hospitals may even provide short-term housing for staff so that they are available to serve patients in need. Local emergency preparedness departments often liaise with other local, state, and federal agencies using pre-established means to share and distribute resources and coordinate responses. The CDC has a Crisis and Emergency Risk Communication (CERC) program that provides training, tools, and resources to help organizations communicate effectively during emergencies.

With regard to clinical engineering's response during events, the main goal is to ensure that there are enough staff and functional equipment to support patient care. In the event of large-scale events, personnel may be asked to perform tasks such as setting up care locations in areas such as lunchrooms and lounges, helping obtain loaner and rental equipment, and any other task that is needed by the hospital.

Infection prevention

As described in Chapter 4, Safety and systems safety, the goal of the infection prevention department (also called the *sterile processing* or *central supply department*) is to prevent the spread of infectious diseases within a healthcare organization. On any given day, about 1 in 25 hospital patients has at least one hospital-acquired infection (CDC, 2018). Many facilities have programs that include the following:

- Enhancing hand hygiene compliance
- Tracking and monitor hospital-acquired infections including, but not limited to
  - catheter-associated urinary tract infections;
  - surgical site infections;
  - hospital-onset *Clostridium difficile* infections;
  - hospital-onset methicillin-resistant *Staphylococcus aureus;*
  - central-line—associated bloodstream infections; and
  - hospital-associated pressure injuries.
- Ensuring proper environmental controls for medication, vaccines, and other biologic materials

Hand hygiene

Hand hygiene means cleansing your hands using either hand washing with soap and water, alcohol-based hand sanitizer, or surgical hand antisepsis. Alcohol-based hand sanitizers are most effective for reducing the number of germs on hands. Antiseptic soaps are the next most effective.

When hands are not visibly dirty, alcohol-based sanitizers are preferred for hand cleaning. Soap and water are recommended when hands are visibly dirty, before eating, after using a restroom, after known or suspected exposure to patients with infectious diarrhea during norovirus outbreaks, and after known or suspected exposure to *C. difficile* (CDC, 2018).

Hospital-acquired infections

The United States is working toward a goal of eliminating hospital-acquired infections. Because of ongoing work, health care is safer now than it was even 10 years ago. In 2019, the Office of Disease Prevention and Health Promotion (ODPHP), part of the US Department of Health and Human Services, published *National Action Plan to Prevent Health Care-Associated Infections: Road Map to Elimination*. This road map set a specific 5-year goal for prevention of hospital-acquired infections. The year 2015 marked the start of new 5-year goals. The CDC leads the United States in tracking hospital-acquired infections and producing the data needed to meet these targets and metrics.

Abbreviations

AAMI Advancement of Medical Instrumentation
AIA American Institute of Architects
ASHE American Society for Health Care Engineering
ASHRAE American Society of Heating, Refrigerating and Air-Conditioning Engineers
CDC Centers for Disease Control and Prevention
CERC Crisis and Emergency Risk Communication
CMS Centers for Medicare and Medicaid Services
DoS denial-of-service
EPA US Environmental Protection Agency
HEPA high-efficiency particulate air
HVAC heating, ventilation, and air-conditioning
MEP mechanical, electrical, and plumbing
NFPA National Fire Protection Association
ODPHP Office of Disease Prevention and Health Promotion
OSHA Occupational Safety and Health Administration
RO reverse osmosis
SDWA Safe Drinking Water Act
USP US Pharmacopeia
US-CERT US Computer Emergency Readiness Team
UVGI ultraviolet germicidal irradiation
WAGD waste anesthetic gas disposal

References

American Institute of Architects. (2009). Design phases. *The architecture student's handbook of professional practice* (14th ed.). Hoboken, NJ: John Wiley & Sons.

American National Standards Institute. (2014). *ANSI/ISEA Z358.1-2014: American National Standard for Emergency Eyewash and Shower Equipment.* Retrieved from <https://webstore.ansi.org/Standards/ISEA/ANSIISEAZ3582014>.

American Society for Health Care Engineering. (n.d.). ASHE Compliance. *Special departmental HVAC issues: operating rooms.* Retrieved from <https://www.ashe.org/topics/regulatory-compliance?page = 6>.

Association for the Advancement of Medical Instrumentation. (2004). *ANSI/AAMI/ISO RD52:2004. Dialysate for hemodialysis.* Arlington, VA: Association for the Advancement of Medical Instrumentation.

Barrick, J. R. & Holdaway, R. G. (2014). *Mechanical systems handbook for health care facilities.* Chicago, IL: American Society for Health Care Engineering.

Centers for Disease Control and Prevention. (2017a). *Developing a water management program to reduce Legionella growth and spread in buildings.* Atlanta, GA: Centers for Disease Control and Prevention.

Centers for Disease Control and Prevention. (February 2017b). Centers for Disease Control and Prevention Healthcare Infection Control Practices Advisory Committee (HICPAC). *Guidelines for environmental infection control in healthcare facilities.* Retrieved from <https://www.cdc.gov/infectioncontrol/guidelines/environmental/index.html>.

Centers for Disease Control and Prevention. (May 3, 2018). *Hand hygiene in healthcare settings.* Retrieved from <https://www.cdc.gov/handhygiene/index.html>.

DiMarco, J. (December 14, 2017). *Medical gas systems: The definitive guide.* Retrieved from Compliant Health Technology <https://www.chthealthcare.com/blog/medical-gas-systems>.

National Fire Protection Association. (2015). *NFPA 99: Health care facilities code.* Quincy, MA: National Fire Protection Association.

National Fire Protection Association (2019). *NFPA 110: Emergency and standby power systems.* Quincy, MA: National Fire Protection Association.

Occupational Safety and Health Administration. (December 2015). *Workplace violence in healthcare.* Retrieved from <https://www.osha.gov/Publications/OSHA3826.pdf>.

Occupational Safety and Health Administration. (January 3, 2019). *Standards intrepretations.* Retrieved from OSHA Standards Interepretations <https://www.osha.gov/laws-regs/standardinterpretations/2009-06-01>.

Office of Disease Prevention and Health Promotion. (January 7, 2019a). *National action plan to prevent health care-associated infections: Road map to elimination.* Retrieved from <https://health.gov/hcq/prevent-hai-action-plan.asp#actionplan_development>.

Office of Disease Prevention and Health Promotion. (January 7, 2019b). *National HAI targets and metrics.* Retrieved from <https://health.gov/hcq/prevent-hai-measures.asp>.

US Computer Emergency Readiness Team. (2019a). *Ransomware.* Retrieved from <https://www.us-cert.gov/Ransomware>.

US Computer Emergency Readiness Team. (2019b). *Security Tip (ST04-015): Understanding denial-of-service attacks.* Retrieved from <https://www.us-cert.gov/ncas/tips/ST04-015>.

US Environmental Protection Agency. (January 12, 2017). *Safe Drinking Water Act (SDWA).* Retrieved from <https://www.epa.gov/sdwa>.

CHAPTER 7

Human resource management

Workforce development, job descriptions, and succession planning

Workforce development

Workforce development attempts to enhance an organization's stability and economic prosperity by focusing on people rather than businesses. It focuses on development of a human resources strategy that addresses issues in matching available workers to needs for the healthcare industry. Workforce development generally focuses on a holistic approach that addresses many different barriers and needs of the healthcare organization.

Successful workforce development programs typically have a strong network of ties in a community and are equipped to respond to changes in their environments. Healthcare entities may have partnerships with educational organizations and provide internships or training pathways to create a pipeline for employees. These partnerships may not only allow access for new employees to enter the workforce, but may also provide opportunities for existing employees to gain additional skills by taking courses or completing degrees to grow and develop their careers.

Within the clinical engineering field, there is an upcoming shortage of staff that has organizations looking toward workforce development programs to ensure that there are enough staff to support the hospital. The 2017 report from *24 × 7 Magazine* shows that the mean age of healthcare technicians increased from 49 to 51 years of age (Gresch, 2018). Gresh claims that a shrinking labor pool has made this aging workforce more complicated. As of June 2018, the unemployment rate was 4.2% nationwide and a low 2.2% in the healthcare and social assistance sector (Gresh, 2018). Compounding this shortage is the recent closing of 33 schools with programs related to healthcare technology management (HTM), leaving only 22 colleges nationwide graduating approximately 400 biomedical equipment technicians (Holt, 2018). These grim statistics have led clinical engineering leaders and hospitals to develop and grow their own biomedical engineering talent through innovative programs (Ruiz, 2018).

Introduction to Clinical Engineering
DOI: https://doi.org/10.1016/B978-0-12-818103-4.00007-7

Job description

A job description is defined as an account of an employee's responsibilities. It usually includes job duties, job responsibilities, and skills needed to perform a role. Tools available from professional societies and peers can assist in the creation of position descriptions that cultivate employee growth and advancement. HTM professionals may lack awareness of career opportunities and how advancement can be obtained. As discussed in Chapter 1, The profession, and in this section, career progression can improve understanding and encourage broad visions for professional futures. As shown in Fig. 7.1, the HTM profession tends to foster specialization and movement into leadership roles.

In addition to the career paths shown in Fig. 7.1, AAMI has developed career progression grids (Tables 7.1 and 7.2) for both technicians and clinical engineers to provide more detail on duties, responsibilities, and skills that can be written into job descriptions (AAMI, 2014).

Table 7.1 lists the AAMI Core Competencies for the HTM-level technician (AAMI, 2016). AAMI created this document to aid academic institutions in the development of curriculum for technical training. It also includes the Medical Engineers and Technician's Association's (META) recommendations for outcomes for BMET programs (META, 2015). Both documents were used to help develop the career progression grids shown in Tables 7.1 and 7.2, which are in turn used to develop job descriptions for members of the field.

Succession planning

Succession planning is the process of identifying and developing personnel within and across an organization so that capable employees are prepared and mentored to cultivate expertise in preparation for the time when existing leaders/advanced technicians leave the organization. Although it seems like the focus of this exercise is "replacement planning," effective succession planning manages the entire talent pool and builds the knowledge, abilities, and skills of staff at every level within the department. Skill development and growth activities also help retain key employees. There are generally five steps in succession planning:

1. Identify key roles for succession planning.
2. Define criteria including skills, knowledge, abilities, and competencies required to undertake those roles.
3. Assess existing staff against the criteria.

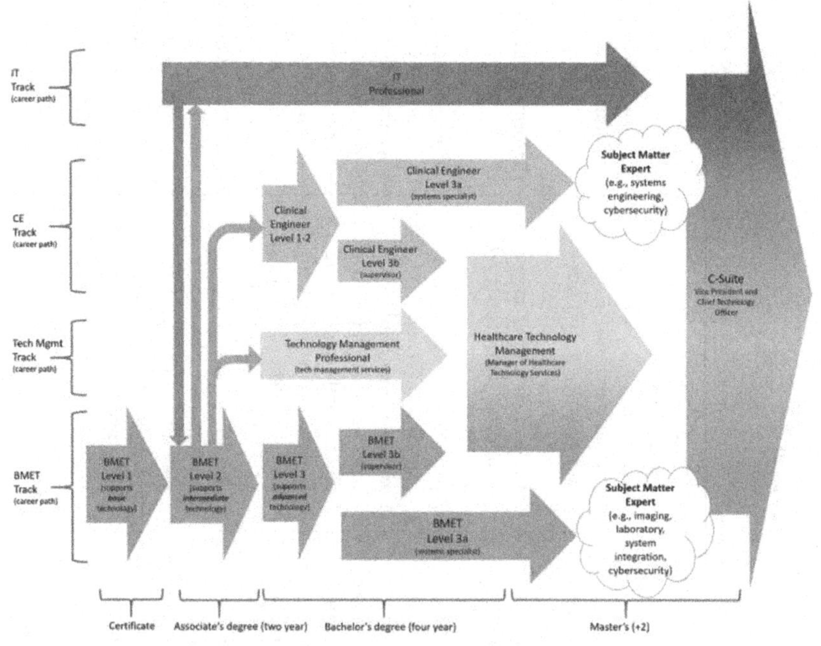

Figure 7.1 Career paths for HTM professionals (Grimes, 2019).

4. Identify possible staff that could perform those criteria.
5. Develop those staff.

With respect to biomedical or clinical engineering departments, one can define key roles within the department as the normal hierarchy of job roles—BMET I, II, and III; specialist; clinical engineer; senior engineer; supervisor; manager; and director. Competencies for each role should be developed (and incorporated into the job descriptions) and development plans for each of these competencies should be developed and provided to the staff identified.

Development tasks should also be incorporated into the goal development discussed in the next section.

Strategic planning, goal development/cascading, and performance evaluations

Introduction

In some organizations, performance evaluations are conducted with no regard or alignment to the organization's larger business needs or goals.

Table 7.1 Technician career progression grid.

Skill/experience	Level I	Level III	Level III	Radiology specialist/network systems specialist/laboratory specialist/project specialist
General Guidelines	Has basic knowledge of job, activity, or function. Needs supervision or mentoring on advanced assignments. Entry level or junior position.	Has comprehensive knowledge and is experienced in most or all facets of job. Has intermediate level of expertise. Capable of assisting less-experienced technicians.	Fully experienced with exceptional skill set or knowledge. Works with minimal supervision. Capable of serving as trainer, mentor to junior and mid-level staff. Capable of performing in lead capacity.	Highly specialized HTM having special training or equivalent in laboratory equipment. Performs highly skilled work of considerable difficulty. Considered technical expert in area of specialty.
Education	Associate degree, military training, or academic work aligned with AAMI Core Competencies (http://my.aami.org/store/detail.aspx?id = CORE-BMET-PDF) and a basic knowledge of mathematics, physics, chemistry, English, and professional skills.	Associate degree, military training, or academic work aligned with AAMI Core Competencies (http://my.aami.org/store/detail.aspx?id = CORE-BMET-PDF) and a basic knowledge of mathematics, physics, chemistry, English, and professional skills, plus additional certification and training as needed.	Associates degree, military training, or academic work aligned with AAMI Core Competencies (http://my.aami.org/store/detail.aspx?id = CORE-BMET-PDF) and a basic knowledge of mathematics, physics, chemistry, English, and professional skills, plus increased levels of certification and training as needed.	Bachelor's degree or associate degree, military training, or academic work aligned with AAMI Core Competencies (http://my.aami.org/store/detail.aspx?id = CORE-BMET-PDF) with substantial experience required, plus increased levels of certification and training in area of specialty.

Leadership	Able to learn from others on the job. Can teach some basic skills to new hires or interns.	Able to learn from others on the job. Can teach basic skills to Level I biomedical equipment technicians (BMETs). Optionally, can mentor others in basic skills.	Adept at learning on the job as well as teaching and mentoring others. Optionally, has developed mastery to the level capable of mentoring other mentors.	Adept at learning on the job as well as teaching and mentoring others. Considered a technical expert in area of specialty and can mentor other mentors.
General Skills and Experience	Has basic understanding and skills related to general electromechanical systems and devices.	Has comprehensive understanding and skills related to general electromechanical systems and devices.	Has advanced understanding and skills related to general electromechanical systems and devices.	Has advanced understanding and skills related to general electromechanical systems and devices as applied to area of specialty.
Specific Experience	• Has basic understanding and can communicate the use of devices supported. • Can provide basic support of acuity equipment for direct patient care. • Is familiar with operations and environment supported such as hospital, clinic, and so on. • Has minimal experience in assigned clinical environment.	• Has comprehensive understanding and can communicate the use of devices supported. • Can provide comprehensive support of acuity equipment for direct patient care. • Is familiar with operations and environment supported such as hospital, clinic, and so on.	• Has detailed understanding and can fully communicate the use of devices supported. • Can provide advanced support of acuity equipment for direct patient care. • Has in-depth understanding of operations and environment supported such as hospital, clinic, and so on.	• Has detailed understanding and can fully communicate the use of devices supported. • Can provide advanced support of acuity equipment for direct patient care. • Has in-depth understanding of operations and environment supported such as hospital, clinic, and so on.

(*Continued*)

Table 7.1 (Continued)

Skill/experience	Level I	Level III	Level III	Radiology specialist/network systems specialist/laboratory specialist/project specialist
	• Has minimal understanding of common clinical terminology and that of life sciences such as anatomy and physiology.	• Has comprehensive experience in assigned clinical environment. • Has comprehensive understanding of common clinical terminology and that of life sciences such as anatomy and physiology.	• Has advanced experience in assigned clinical environment. • Has in-depth understanding of common clinical terminology and that of life sciences such as anatomy and physiology.	• Has advanced experience in assigned clinical environment. • Has in-depth understanding of common clinical terminology and that of life sciences such as anatomy and physiology.
Public Safety and Regulatory Requirements	Has basic understanding of both local and national public safety and regulatory issues.	Has comprehensive understanding of both local and national public safety and regulatory issues.	Is knowledgeable about both local and national public safety and regulatory issues.	Is knowledgeable about both local and national public safety and regulatory issues, especially those that apply to area of specialty.
Customer Service	Can solve basic frontline customer service issues.	Can solve service-line customer service issues.	Can successfully solve organization-level customer service issues and complaints.	Can successfully support solution of organization-level customer service issues and complaints, especially those that apply to area of specialty.

Specific Equipment Expertise	• Has basic understanding of clinical equipment such as radiologic, laboratory, and networked medical systems. • Has basic understanding of project management terms and methods.	• Has working knowledge of clinical equipment such as radiologic, laboratory, and networked medical systems. • Has basic understanding of project management terms and methods.	• Has working knowledge of clinical equipment such as radiologic, laboratory, and networked medical systems so that work beyond single devices can be appropriately accomplished. • Has basic understanding of project management terms and methods.	• Has working knowledge of clinical equipment such as radiologic, laboratory, and networked medical systems so that work beyond single devices can be appropriately accomplished. • Has basic understanding of project management terms and methods. Project Specialists have advanced mastery and certification in project management methodologies.

From Association for the Advancement of Medical Instrumentation. (2014). *AAMI career planning handbook*. Retrieved from <http://www.aami.org/membershipcommunity/content.aspx?itemnumber = 1485&navItemNumber = 787>.

Table 7.2 Clinical engineering career progression grid.

Skill/Experience	Staff Engineer	Networked Systems/integration engineer	Senior Engineer
General guidelines	Has basic knowledge of job, activity, or function. Needs supervision or mentoring on advanced assignments.	Has basic knowledge of job, activity, or function. Needs supervision or mentoring on advanced assignments.	Fully experienced with exceptional skill set or knowledge. Works with minimal supervision. Capable of serving as trainer, mentor to junior and mid-level staff. Capable of performing in lead capacity.
Education	Bachelor's degree in engineering in related discipline required. Master's degree desired.	Bachelor's degree in engineering in related discipline required. Master's degree desired.	Bachelor's degree in engineering in related discipline required. Master's degree desired.
Leadership	Able to learn from others on job. Can teach basic skills to entry-level staff. Optionally, can mentor others in basic skills.	Able to learn from others on job. Can teach basic skills to entry-level staff. Optionally, can mentor others in basic skills.	Adept at learning on job as well as teaching and mentoring others. Desirable to be considered technical expert and able to mentor other mentors.
General Skills and Experience	Has comprehensive understanding and skills related to general electromechanical systems and devices.	Has comprehensive understanding and skills related to general electromechanical systems and devices.	Has advanced understanding and skills related to general electromechanical systems and devices.
Specific Experience	• Is familiar with operations and environment supported such as hospital, clinic, etc. • Has minimal experience in assigned clinical environment.	• Has comprehensive understanding of operations and environment supported such as hospital, clinic, etc.	• Has advanced understanding of operations and environment supported such as hospital, clinic, etc.

	• Has basic understanding of common clinical terminology and that of life sciences such as anatomy and physiology.	• Has comprehensive experience in assigned clinical environment. • Has minimal understanding of common clinical terminology and that of life sciences such as anatomy and physiology.	• Has advanced experience in assigned clinical environment. • Has basic understanding of common clinical terminology and that of life sciences such as anatomy and physiology.
Public Safety and Regulatory Requirements	Has basic understanding of both local and national public safety and regulatory issues.	Has basic understanding of both local and national public safety and regulatory issues.	Is knowledgeable about both local and national public safety and regulatory issues.
Customer Service	Can successfully solve organization-level customer service issues and complaints.	Can successfully solve organization-level customer service issues and complaints.	Can successfully support solution of organization-level customer service issues and complaints, especially those that apply to area of specialty.
Specific Equipment Expertise	• Has basic understanding of clinical equipment such as radiological, laboratory, and networked medical systems. • Has basic understanding of project management terminology and methodology.	• Has basic knowledge of clinical equipment such as radiological and laboratory devices. • Has mastery of networked medical systems so that work beyond single devices can be appropriately accomplished. • Has basic understanding of project management terminology and methodology.	• Has mastery of general medical surgical equipment. • Has working knowledge of clinical equipment such as radiological, laboratory, and networked medical systems. • Has mastery of project management terminology and methodology.

From Association for the Advancement of Medical Instrumentation. (2014). *AAMI career planning handbook*. Retrieved from <http://www.aami.org/membershipcommunity/content.aspx?itemnumber = 1485&navItemNumber = 787>.

Figure 7.2 Cyclical nature of strategic and departmental goal development, cascading, and measurement.

However, this misalignment can limit the growth and development of organizations and individuals. Therefore, larger, more advanced organizations will align strategic plans with individual and departmental goals. This section will cover developing and aligning goals. The next sections will discuss aligning these goals with performance evaluations and strategic planning. Fig. 7.2 depicts the cyclical nature of strategic planning; organizational, department, and staff-level goal development; and performance and goal evaluations. In general, organizations will complete an extensive strategic plan every 5 to 10 years. Clinical engineering departments should use this organizational strategic plan as their own departmental strategy every 3 to 5 years. It is from these departmental strategic plans that yearly goals can be developed and cascaded to staff. These goals are annually evaluated at the staff and department level and will inform the next round of strategic planning.

Strategic planning

Strategic planning is the process of defining an organization or department's direction and making decisions on the allocation of resources to pursue this strategy (Allison & Kaye, 2005). There are many books that provide methodologies to develop a strategic plan, but development of a strategy generally means setting goals, determining the actions or tactics that will be used to meet these goals, and then allocating resources (e.g., people, money) to complete the actions. Strategies are usually developed for longer terms (3 to 10 years) but can be aligned with monthly, quarterly, and annual planning processes that occur within an institution. Planning is paramount as resource allocation through the operating and capital budgets can be on different timelines as the organizational planning process.

To be successful, strategies also need to be tied to the mission, vision, and values of the organization. Collis and Rukstad provide a clear hierarchy of company statements that help determine the difference between mission, values, and vision statements (Collis & Rukstad, 2008). This hierarchy is illustrated in Fig. 7.3. Additionally, it shows that a strategy cannot be successful unless you measure and monitor the strategy using a balanced scorecard or other measurement mechanism. Ensuring that the goals and tactics used in a strategic plan align to the mission, values, and vision of the organization ensures that the plans will be supported by the organization, so they are more likely to be successful. Measuring these goals and tactics is now standard practice, so writing goals that can be measured is paramount.

Normally, organizational-level strategic planning occurs every 5 to 10 years. Once organizational direction is set, clinical engineering department leaders should conduct their own strategic planning sessions to set their

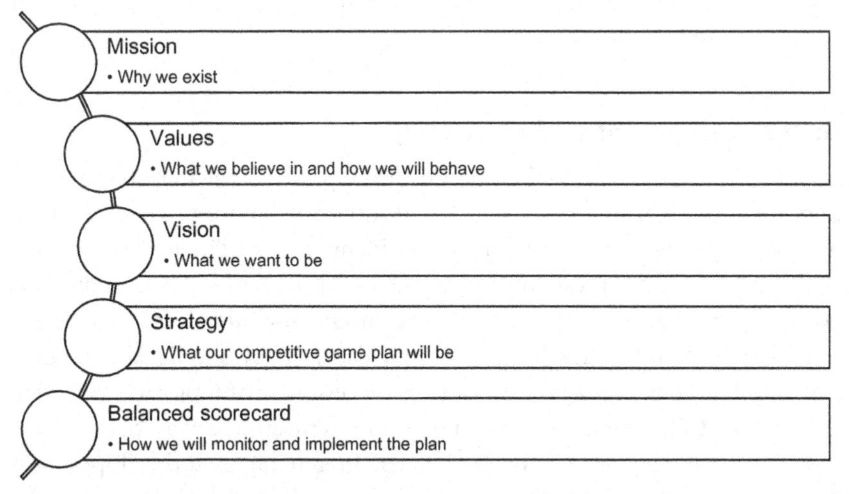

Figure 7.3 A hierarchy of company statements. *Modified from Collis, D.J., & Rukstad, M.G. (April 2008). Can you say what your strategy is? Harvard Business Review, 1-9.*

Table 7.3 Template for a strategy map for a department that is aligned to organization mission, vision, and values.

Mission/vision/ values statements	Mission #1	Mission #2	Mission #3
Strategies	Strategies to meet Mission #1	Strategies to meet Mission #2	Strategies to meet Mission #2
Tactics	Tactics to meet strategies	Tactics to meet strategies	Tactics to meet strategies
	Measurements to show implementation/ success of tactics	Measurements to show implementation/ success of tactics	Measurements to show implementation/ success of tactics

departmental vision for the next 3 to 5 years. These departmental strategies and tactics should align to the organizational strategy and set a direction for the clinical engineering department to grow and develop.

Once strategic goals are defined (at either the organizational or departmental level), leadership can develop a one-page diagram to document and communicate its strategy to team members and other leaders within that organization. Table 7.3 shows a template for a strategy map. Mission/vision/values statements are listed in columns, and the strategies or goals defined by the department are listed in the column that aligns to the mission. Below that, tactics and measurements for each strategy are outlined. In this configuration, all measurements, tactics, and strategies are aligned to the organization in a clear column. A new row is inserted for each tactic and measurement.

Goal development and cascading

At least annually, departments should develop goals that help them reach their departmental strategic plan. There should be approximately three to five departmental goals annually to maintain focus. These departmental goals can be broken down (or cascaded) into team and/or individual goals. When creating these goals, they should include short- and long-term activities and defined start and end dates. They should also include any development plans created during succession planning or workforce development planning. Goals should also be written in a way that is measurable to make evaluating the goal easier. In general, goal development is done with the staff members so that they are aligned and buy in to the tasks/goals assigned to them.

Performance evaluations

A performance evaluation is a systematic process that assesses an individual employee's performance as measured against pre-established objectives and goals. This assessment is usually evaluated on a regular basis (either annually or on the employee's service anniversary). Evaluations can be used in making pay increase decisions and other personnel decisions including promotions and demotions.

Evaluations may have multiple sections including appraisals of the following:

- *Behaviors or core values the organization has set.* These behaviors and values are measures of "how" staff achieve their goals and may include behaviors such as respect, integrity, and teamwork.
- *Role responsibilities.* These responsibilities are based on the staff members' job description and are sometimes described as *job competencies.*
- *Individual goals.* These goals are developed in conjunction with the employee's leader, usually at the beginning of the review period, and are generally linked to the needs of the department and organization.

Organizations often link their goals starting from their strategic plan, down through the executive level to the frontline staff so that all staff are aligned in achieving the often complex goals to improve and enhance a hospital or health system. Strategic plans are often in place for 3 to 5 years, which allows for ample time to align annual goals to achieve these plans.

Annual performance evaluations are also a time for staff to discuss their ongoing learning and development needs with leadership. Evaluations offer an opportunity for staff to discuss their growth plans and discuss needs for additional training and development opportunities to continue to learn and grow within the organization.

Once individual staff performance evaluations are complete, departmental leadership should also evaluate the departmental goals. This assessment will inform the departmental leadership in their goal setting for the next year's activities. Once the cycle is complete, it begins again with another year of goal planning and cascading these goals.

Change management

Change management is the controlled identification and implementation of a change within an organization. It has grown into an important field that supports all types of changes including organizational, departmental, team, and individuals.

Change management models

There are many different change management models. In healthcare, the two most frequently used are *Plan-Do-Check-Act (PDCA)* created by W. Edwards Deming, and *Leading Change*, an eight-step model created by Dr. John Kotter. The PDCA model focuses on continuous improvement and is usually applied to frontline staff. The Leading Change model is generally focused on larger organizational changes and is usually focused on hospital or health center leadership levels.

Plan-Do-Check-Act

PDCA is a continuous quality improvement process used frequently by hospital and health system quality departments to implement smaller process changes within departments or across the system (Tague, 2005). This process is frequently used because it allows for small changes to be implemented, usually in a quick-turnaround methodology, which allows for fast improvements that can be implemented by frontline staff. Fig. 7.4 shows the PDCA process as a continuous loop.

Plan

The plan phase requires evaluating a current process within or across departments and determining if and how it can be improved. Ideas are generated, and small changes are planned. Usually only small changes are planned so that expected outcomes from the change can be defined and measurements of the affected change can be delineated. Small changes

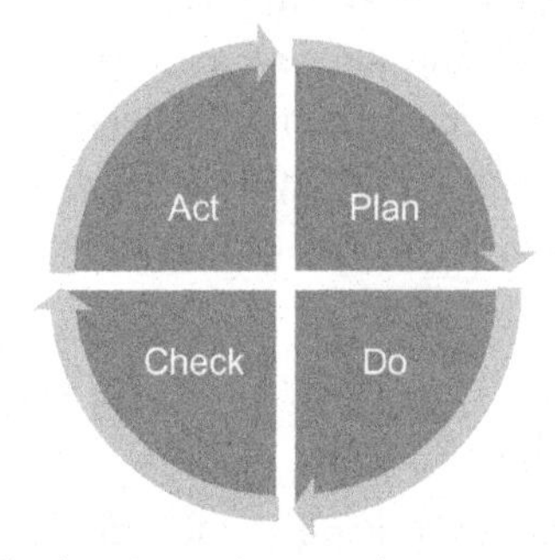

Figure 7.4 The Plan-Do-Check-Act continuous change management process.

allow for a stepwise continuous change process that creates a cause-and-effect mechanism that can be measured and tracked.

Do

The do phase implements a single change from the plan phase.

Check

During the check phase, the measurements, data, and results are collected from the do phase. Results are compared with the expected outcomes to determine if the change had the desired effect. The measurements are also assessed to determine if they are effective. Generally results are charted against time to determine if the change is having the desired effect. These charts can be continued as different changes are made to view the long-term impact on the outcome as an effect of several different implemented changes.

Act

If the data in the check phase show that the change was beneficial to the organization and met the goal for improvement, then the change can be implemented permanently. If the change was not successful, the existing process would remain in place. Regardless of whether or not the change was successful, additional ideas for improvements can be generated in this phase to continue to enhance the process. These further ideas for improvements are then recycled into the process as the plan step is begun again.

Kotter's Leading Change eight-step management process

Dr. John Kotter, the Konosuke Matsushita Professor of Leadership, Emeritus at Harvard Business School, created the Leading Change eight-step management process (Kotter, 1996). This process is not specific to healthcare but has been taught by human resources departments in hospitals and health systems as the process is easily applicable to many situations within the healthcare arena. Kotter's original book was published in 1996 and updated in 2012. The eight steps of this process include:

- creating a sense of urgency;
- building a guiding coalition;
- forming a strategic vision and initiatives;
- enlisting a volunteer army;
- enabling action by removing barriers;

- generating short-term wins;
- sustaining acceleration; and
- instituting change.

This change method is usually taught to leaders within the organization as the steps include the requirement for leaders to complete tasks such as building a coalition and removing barriers. Generally these tasks cannot be completed by frontline staff alone and will require some leadership buy-in and intervention.

Reacting to change

When implementing changes, not only do leaders have to plan how to implement the change, they also have to be prepared for others' reactions to the proposed change (Heller, 1998). Robert Heller proposed several steps people go through as they are presented with a change. These are:

- stability at the point of change (the current state before the change was introduced);
- inability to act (the initial phase when presented with the change);
- denial;
- anger;
- bargaining;
- depression;
- testing; and
- acceptance.

This cyclical nature requires astute leaders to understand where staff are in this cycle and support them as they move through these steps to ensure that the change is not only implemented successfully but also is sustained. When introducing a change to their staff, leaders must remember that they have already had time to question, think about, understand, and accept the change. However, often they present changes to staff with little or no time allotted for staff to go through those same steps of acceptance and understanding. By properly preparing and anticipating staff's reactions to the change, as well as allowing the staff time to question, understand, and accept the change, leaders can introduce and adopt changes much more successfully.

Culture of safety

In 1999, the Institute of Medicine published *To Err is Human: Building a Safer Health System*, a report that highlighted medical errors at an incidence of 44,000 to 98,000 deaths annually (Institute of Medicine et al., 2000) To put this into perspective, this would equal a plane crash killing everyone on board every single day of the year. Even though thousands of lives are saved every day at healthcare institutions, patients are also harmed every day.

To Err is Human was one of the first reports that introduced medical error as a cause of death, however the data were based on primary research. A more recent study claims that more than 251,000 people in the United States die each year from medical errors (Makary & Daniel, 2016). This now makes medical error the third leading cause of death in the United States behind cancer and heart disease. Both studies have spurred the healthcare industry to move toward practices and tools that make its practice safer.

The American Nurses Association (ANA) describes a "culture of safety" as the core values and behaviors that come about when organizational leadership, managers, and healthcare workers have a collective and continuous commitment to emphasize safety over competing goals (ANA, 2019).

The concept of safety culture originated outside healthcare. Studies of high-reliability organizations, such as nuclear power and avionics, show organizations that consistently minimize adverse events despite carrying out intrinsically complex and hazardous work. High-reliability organizations maintain a commitment to safety at all levels, from frontline providers to managers and executives. This commitment establishes a "culture of safety" that encompasses the following key features:

- Acknowledgment of the high-risk nature of an organization's activities and the determination to achieve consistently safe operations.
- A blame-free environment where individuals are able to report errors or near-misses without fear of reprimand or punishment.
- Encouragement of collaboration across ranks and disciplines to seek solutions to patient safety problems.
- Organizational commitment of resources to address safety concerns (Patient Safety Network, 2019).

Organizations can measure their culture of safety using validated surveys including the Agency for Healthcare Research and Quality (AHRQ) Survey on Patient Safety Culture and the Safety Attitudes Questionnaire. The AHRQ also provides yearly updated benchmarking data so hospitals and healthcare systems can measure where they are compared with their peers.

Hospitals and healthcare entities use various tools and techniques to improve their culture of safety including error-prevention training.

Error prevention

Error-prevention techniques and tools were designed to reduce the majority of errors seen in healthcare settings. Humans experience three different types of errors: skill-based errors, rule-based errors, and knowledge-based errors (Embrey, 2005). The name of the error type describes the mode that an individual's brain is in at the time he or she experiences the error. Each mode represents a different level of familiarity with the task being performed and the degree of conscious thought that is applied when performing the task.

Skill-based errors

When an individual is in skill-based performance mode, a well-developed skill pattern exists that has been developed through practice and repetition of an act. These are routine and familiar tasks performed when one does not even have to think about it (Embrey, 2005). Examples include brushing your teeth and driving home. Skill-based errors are ones in which an individual does something unintended, even though he or she knows how to do it. An example is sending an email without the attachment.

Rule-based errors

When individuals are in rule-based performance mode, they are perceiving a situation and the brain will scan through all past experiences and education to find a rule they know. They then try to apply the rule to the act or situation they are in. When making a rule-based error, individuals may either misapply the rule because it does not apply to the situation or choose not to comply with the rule (Embrey, 2005). For example, we know we are not supposed to drive over the speed limit, but we do.

Knowledge-based errors

When people are in a new or unfamiliar situation or have not yet developed a skill, they enter a problem-solving mode. In this mode, they try to "figure it out." Decisions are made very slowly because of the work to "figure it out," and they are very error-prone because they may rely on a good memory or a guess (Embrey, 2005). For example, if asked to calculate the volume of a cone 3 m in diameter and 4 m tall, most people would take a long time to try to figure it out, and the answers would not be very accurate. They would rely on information possibly learned a long time ago and not frequently used.

By applying different tools, hospitals have begun teaching their staff ways of minimizing or eliminating skill-based, rule-based, and knowledge-based errors. The following sections provide some examples of these tools.

STAR

STAR is a simple, four-step technique to improve attention to detail. The best times to use STAR are when staff are going from thought to action, such as medication administration, entering data into a device or computer, sending an email with an attachment, or adding a value on a spreadsheet.

The STAR technique is used for preventing skill-based errors. STAR is an acronym for:

- *Stop*. Pause for 1 to 2 seconds to focus attention on the task at hand.
- *Think*. Visualize the act, and think about what is to be done.
- *Act*. Concentrate and perform the task.
- *Review*. Check for the desired result.

Three-way repeat back

The three-way repeat back is a powerful tool for clarifying communication. Sometimes people need the information conveyed differently than how it is being presented. In healthcare, it is important that all parties understand and clearly reiterate the information in that moment. This tool helps clarify the information by having the person receiving the information validate the information provided. The steps for the three-way repeat back include the following:

1. The sender initiates communication using the receiver's name and then provides an order, request, or information to the receiver in a clear, concise format.
2. The receiver acknowledges receipt by a repeat back of the order, request, or information.
3. The sender acknowledges the accuracy of the repeat back or repeats the communication if it is not accurate.

SBAR

SBAR is used to hand off when communicating about a problem or issue that needs resolution. This communication option offers more detail than a three-way repeat back and provides an opportunity for the individual to deliver his or her recommendation. SBAR is an acronym that stands for the following:

- *Situation.* Identify who and what as well as the immediate problem, concern, or issue.
- *Background.* Review pertinent information, such as environment, procedures, and employee status.
- *Assessment.* Offer your view of the situation or urgency of action. For example, you can say, "I think the problem is...," "I'm not sure what the problem is...," or "The situation is deteriorating rapidly, and we need to do something."
- *Recommendation.* Offer your suggestion or request to the other person.

ARCC

ARCC is an error-prevention tool that is used to help escalate a concern in a nonthreatening way. ARCC helps advance the concern if it is not addressed in a respectful way. It can be especially helpful if an individual feels hesitant or intimidated to raise a concern to someone he or she perceives to be in a position of higher authority. One may not need all the steps to obtain the desired results. Here is how it works:

- *Ask* a Question;
- Make a *Request*;
- Voice a *Concern*;
- Use *Chain of Command*.

Stop the line

This error-prevention tool is used to stop any action and reassess the situation so that everyone is on the same page before continuing. Stopping the line should always be done in a calm and respectful voice. Here's how it works. Simply say: "Please stop the line, I need clarity." Anyone in the organization has the authority to STOP THE LINE any time that an immediate threat (real or perceived) to patient or staff safety is identified.

Although these tools are some of the most commonly used, many other techniques and tools are being used to further enhance patient and employee safety and reduce the incidence of medical errors. The AHRQ has may other techniques such as Communication and Optimal Resolution (CANDOR), which provides a process to respond immediately when patients are harmed (AHRQ, 2014). AHRQ also provides a toolkit for completing family-centered rounds and a risk assessment toolkit to use for facility design safety risks (AHRQ, 2014). Different healthcare entities may choose which tools and strategies to implement to help them reduce their overall medical error rate.

Project management

Project management is the discipline of planning and organizing resources to achieve a goal. Projects are temporary endeavors that have a defined beginning and end and are generally constrained by either funding, personnel (resources), or time. Projects generally bring about a beneficial change or effort or an added value to the organization. There are many types of projects undertaken by hospitals that clinical engineering departments may take part in (as a resource) or may lead including, but not limited to the following:

- *Improvement projects*. These are projects with the goal to improve upon a process or outcome.
- *Expansion projects*. These projects expand current hospital operations including new locations and new construction.
- *Implementation projects*. These are projects with the goal to implement a new piece of equipment or technology.

There are several different approaches to project management that use different techniques and tools to complete the project. The Project Management Institute (PMI), a nonprofit professional organization for project management, provides credentialing and certification for project managers using their standard methodology. *Lean project management* uses lean principles applied to project management concepts. *Critical chain project management* uses a methodology that puts an emphasis on the resources needed to complete a project. *Extreme project management* is a project management methodology that is very complex and very uncertain.

Most project management methodologies use a similar set of phases to identify the process through which the projects move. These phases generally include:

1. initiation;
2. planning and design;
3. execution/construction and testing;
4. monitoring and controlling; and
5. close-out or completion, and transition to daily operations.

During these phases, a project team is created. This team generally includes an executive sponsor, a project manager, project stakeholders who are affected by the project or have an interest in its completion, and resources to complete the project work. The project team creates a project charter and budget, when appropriate. A project timeline is created with tasks that need to be accomplished, and resources are assigned to these tasks.

The project manager, sometimes titled the *project coordinator* or *technical manager*, ensures that the project group maintains focus, progresses appropriately, and meets the deliverables in a timely way. Communication excellence is a cornerstone of the project manager's duties. A good manager determines how much information each stakeholder needs and when it is communicated. Scheduling is managed by the team leader. In addition, tasks include the management of activities and challenges that arise. Group meetings are led by the project manager. Conflicts between team members are resolved by the team leader. Lastly, detailed and effective documentation of project activities and outcomes is a crucial part of project management. Project manager skills are multifaceted and summarized as:

- leadership generating collaboration among team members;
- organizational skills and attention to detail; establishment of priorities;
- people management, including delegation of tasks, progress monitoring, and team building;
- communication, including both written and verbal tools to convey information to stakeholders with a variety of backgrounds;
- time management, meeting deadlines, and dealing with setbacks; and
- technical expertise to understand the overarching project goals.

Abbreviations

AHRQ Agency for Healthcare Research and Quality
BMET Biomedical equipment technician
CANDOR Communication and Optimal Resolution
HTM Healthcare technology management
META Medical Engineers and Technician's Association
PDCA Plan-Do-Check-Act
PMI Project Management Institute
ANA The American Nurses Association

References

Agency for Healthcare Research and Quality (2014). *AHRQ Quality Indicators™ toolkit for hospitals*. Retrieved from https://www.ahrq.gov/patient-safety/settings/hospital/resource/qitool/index.html.

Association for the Advancement of Medical Instrumentation. (2014). *AAMI career planning handbook*. Retrieved from <http://www.aami.org/membershipcommunity/content.aspx?itemnumber = 1485&navItemNumber = 787>.

Association for the Advancement of Medical Instrumentation. (2016). *Core competencies for the HTM entry-level technician. A guide for curriculum development in academic institutions*. Retrieved from <http://www.aami.org/productspublications/ProductDetail.aspx?ItemNumber = 924>.

Allison, M., & Kaye, J. (2005). Strategic planning for nonprofit organizations. A practical guide and workbook (2nd ed.). Hoboken, NJ: John Wiley & Sons.

American Nurses Association. (2019). *Culture of safety*. Retrieved from <https://www.nursingworld.org/practice-policy/work-environment/health-safety/culture-of-safety/>.

Collis, D. J., & Rukstad, M. G. (April 2008). Can you say what your strategy is? *Harvard Business Review*, 1–9.

Embrey, D. (2005). *Understanding human behaviour and error. Human error*. Human Reliability Associates. Retrieved from <http://www.humanreliability.com/downloads/Understanding-Human-Behaviour-and-Error.pdf>.

Grimes, S. (2019). CE/HTM professional roles in healthcare delivery. *Biomedical Instrumentation and Technology*, *53*(3), 206.

Gresch, A. (June 25, 2018). *Overcoming the Biomed tech shortage*. Retrieved from <http://www.24x7mag.com/2018/06/overcoming-biomed-tech-shortage/>.

Heller, R., & Hindle, T. (1998). *Essential managers: Managing change*. New York, NY: DK Publishing.

Holt, C. (August 17, 2018). *Confronting the BMET staffing shortage*. Retrieved from <http://www.24x7mag.com/2018/08/confronting-bmet-staffing-shortage/>.

Institute of Medicine, Committee on Quality of Health Care in America, Kohn, L. T., Corrigan, J. M., & Donaldson, M. S. (Eds.), (2000). *To err is human: building a safer health system*. Washington DC: National Academies Press.

Kotter, J. (1996). *Leading change*. Boston, MA: Harvard Business Review Press.

Makary, M. A., & Daniel, M. (2016). Medical error—the third leading cause of death in the US. *BMJ*, *353*, i2139. Available from https://doi.org/10.1136/bmj.i2139.

Medical Engineers and Technician's Association. (2015). *META outcomes*. Retrieved from <www.mymeta.org>.

Patient Safety Network. (2019). *Culture of safety*. Retrieved from <https://psnet.ahrq.gov/primers/primer/5/culture-of-safety>.

Ruiz, J. (July 3, 2018). *Growing a career: Cultivating your own biomed tech talent*. Retrieved from <http://www.24x7mag.com/2018/07/cultivating-biomed-tech-talent/>.

Tague, N. R. (2005). *The quality toolbox* (2nd ed.). Milwaukee, WI: ASQ Quality Press.

Appendix: Additional Readings

Article 1: An Assessment of industrial activity in biomedical engineering. National Academy of Engineering (1971)

Introduction

The National Institutes of Health (NIH) is a leading force in the conduct of the national research program in medicine and biology. A better definition of the factors relevant to industrial participation in the field of biomedical engineering is vital to them if they are to chart appropriate courses of future action. Thus, in 1965 the Resources Analysis Branch of NIH sponsored a pilot study[3] that explored methods by which NIH could gather and analyze data pertinent to industrial support of biomedical research and development. This study was performed by Arthur D. Little, Inc. It included interviews with 17 industrial companies, resulting in the development of a proposed questionnaire and methodology to be used as a primary data gathering instrument by the government.

For several reasons, the survey that might have followed this proposed design was not implemented. However, still recognizing the need for the information that it might have revealed, in June 1969, the NIH Office of Resources Analysis asked the National Academy of Engineering to conduct an assessment of current attitudes and commitments of industry in the area of biomedical engineering. The statement of work in the contract between NIH and the Academy asked the Committee on the Interplay of Engineering with Biology and Medicine (CIEBM) to:

1. Conduct in-depth surveys involving key personnel of 50 selected companies.
2. Identify industrial activity, capabilities, and plans with respect to all areas of biomedical engineering.
3. Establish guidelines for the development of the national industrial biomedical effort.
4. Provide recommendations to the National Institutes of Health in furthering this effort.

To implement this task, the CIEBM constituted the Task Group on Industrial Activity, which immediately initiated the effort. The study was an integral part of the overall program of the CIEBM. (Appendix A provides a brief description of other committee involvements in bioengineering and how they relate to the industrial assessment reported herein.)

The conclusions and recommendations of the Task Group reflects a careful evaluation of the data collected during this study. Also, the Task Group was strongly influenced by the extensive experience of its members and the knowledge gleaned from numerous involvements in biomedical engineering.

The following chapters of this report on industrial activity in biomedical engineering present the conclusions and recommendations of the Task Group, describe the procedures used in the investigation, and discuss the findings derived therefrom.

Summary and Conclusions

Four major and broad factors that affect industrial participation in biomedical engineering were identified during this study: communication between those who must interact in the field; the nature of the marketplace in which industry must operate; government involvement in the various aspects of biomedical engineering; and product efficacy and safety. Clearly these factors and the problems they introduce are interrelated; to isolate one for solution without giving consideration to all would result in little headway toward achieving greater national benefits from biomedical engineering products and services. Each of the four factors are discussed more completely below together with major, general conclusions that are derived therefrom. This chapter ends with a concise list of specific conclusions.

Communication between and within the Medical and Technological Communities

There are two processes – mutual acceptance of concepts and identification of appropriate roles – that are essential to an effective interaction between the engineering and medical professions. Both involve communication, and both are inadequate at present. This is evident from the standpoint of the biomedical engineering industry and the engineers in its employ as well as from the standpoint of the medical profession and the personnel responsible for the application of technological products to patient care.

While perhaps not as formidable as some have characterized it in the past, this communication barrier still impedes a systematic approach to the solution of problems that require the combined talents of the physical scientist, the engineer, the life scientist and the physician. The basis of understanding and acceptance between the medical and engineering professions pertaining to the nature of biomedical equipment and its application, which would permit widespread and thorough collaboration and acceptance among these disciplines, does not yet exist.

To overcome these barriers, a larger common ground of knowledge and understanding is required. This implies the need for an alteration and augmentation of the educational process in which knowledge is acquired by all of those involved in health care – the engineer and the physician, the technician and the nurse, the hospital administrator and the serviceman. It is a joint responsibility of the engineering school, the medical school, industry, and the government. *The potential of large-scale educational programs at all levels to promote knowledge and to create an atmosphere of mutual respect and acceptance of ideas among the professions is very great. It should not be ignored as we develop a systematic approach toward seeking solutions to vital problems in health care.*

The Biomedical Engineering Marketplace and Industrial Capability

The second factor involves the manner in which industry operates and the utilization of its many capabilities. The biomedical engineering industry, like other industrial endeavors, operates under certain constraints. First, a reasonable profit is required. Industry has demonstrated its interest and ability to assume development and productive initiative when an adequate return on its investment is envisioned. Second, a responsible industrial corporation has a tendency to limit its products and services to those that use well-established technology indigenous to the company. This tendency can limit the expansion of such emerging technological application areas as biomedical engineering.

In addition to these common factors, however, there are others that combine to make the diverse and individualistic biomedical engineering marketplace greatly different from that usually encountered by United States industry; these factors tend to inhibit the investment of private-risk capital and resources.

One difficulty faced by industry is that of determining the market potential. While companies seem to be able to estimate the size of the marketplace for established product lines, even the largest of corporations find it difficult to predict the market for new devices or services in this biomedical field. Many market failures resulting from this situation have contributed to the conservative attitude within industry, particularly when risk investments are being considered.

Another problem is the unorganized manner in which product ideas and productive capacity are brought together. Numerous instances can be found where physicians and researchers have a particular need for a product but at the same time have no knowledge of where to locate a source for its development. Similarly, there are many industries that have the capability to produce specialized products, but lack the knowledge of an existing need and market. Major sources of product ideas are conventions, published research papers, and the advice of one or, at most, a few medical consultants.

A third problem results from the fact that certain devices, while highly desirable, need only to be produced in small quantities. The problem is one of finding a source for development of such products on a practical basis. Large corporations, those having the strongest position in the biomedical industry, are not the solution to this problem because of the extensive overhead that they carry. There are many small, highly qualified, and technically oriented firms that have the capability and, because of their limited overhead, are able to tackle such problems. When coupled with the convictions of one or two key individuals, such companies are willing to and do operate in this field. There is an urgent need to bring such companies together with the medical community. However, a liaison function is not the entire solution, for while the small firm possesses the technological capability, it may lack not only a sufficient knowledge of its potential market but, also, the knowledge of how to approach this market. Some of this is due to a lack of experience. Primarily, however, it is due to a lack of the funds required to perform adequately these marketing functions. This is one area where government involvement can provide an important impetus to product development and market introduction. Most large corporations would prefer to pursue the marketing problems on their own, perhaps with some assistance to overcome development problems, but the small firm requires guidance in all three phases – product development, market identification, and the actual marketing program.

It must further be recognized that another class of products is not attractive to any industry, large or small. This class is composed of those products that, while having high value in medical care, require development expenditures that cannot be realistically recouped from the limited market for them. Such products will require direct subsidy if private industrial capability is to be utilized in their development, production, and deployment:

Another market difficulty results from the manner in which products achieve user acceptance. The process is a long one due to the need for careful clinical evaluation before marketing and the justifiable conservative attitude of the medical profession when patient welfare is involved. This leads to longer time periods from product inception to profitable sales than those encountered in many other product lines, thus delaying the return on investment.

In summary, the Task Group believes that these conditions have combined to retard the utilization of the available capabilities of our industrial resources. The productive output is considerably below that which could be achieved, and the biomedical engineering products and services that are provided do not fully reflect recent advances in technology.

Of all of the factors that deleteriously affect industrial biomedical engineering activity, the lack of clear economic incentives, which exists as a result of the present nature of the marketplace, is seen by both industry and the

Task Group to be the greatest single impediment to beneficial expansion of industrial efforts. *The full potential of United States industry has not been realized in the field of biomedical engineering. To harness this capability, a viable marketplace needs to be developed.*

Government Involvement in Biomedical Engineering Development

Interviews with industry during this study and discussions with industry in the workshop following this study strongly supported the Task Group's belief that there is a lack of understanding within industry as to the position of the government in its involvement with the national biomedical engineering effort. This pertains to the areas of funding, patent regulation, and the role of NIH and other agencies in development activities.

In particular, industry (and others) often cannot identify the responsibilities of and the relationships among the various government agencies involved in health care. The tendency is to think of NIH as the focus, even though other agencies (e.g., National Center for Health Services Research and Development) may more properly be responsible for a specific activity. The result is that industry has difficulty in locating within the government the proper sources of advice and counsel, technical and market information, and funds needed to pursue important problems.

A related problem is that of differing policies among various government departments. One problem examined in depth in a workshop[4] sponsored by the Academy in 1969 and raised again during the industrial study is patent policy. Industry perceives HEW as having a patent policy prejudiced against profit-making concerns, a policy quite different from that of the DOD and NASA with whom technological companies have had much experience. Whether or not this industrial perception has a basis in fact, free enterprise is stifled when a firm is of the opinion that it will not be able to maintain the rights to a development in which it will invest considerable expenditures in development and market establishment. Perhaps full patent rights are not justified in some instances where considerable government subsidy is provided; however, consideration should be given to the fact that a firm needs sufficient time to assure marketability and acceptance of its product, to attain a reasonable return on its investment in research and development, manufacturing, and marketing efforts. The need for the formulation and dissemination of a uniform but flexible government policy in this area is great.

Because of the scope of these problems and the extensive effort required for their solution, *there has been and continues to be a strong need for government involvement to assure that the problems do reach a solution. This involvement takes the form of both financial assistance and organizational direction.*

Provision of Adequate Consumer Protection

The question of medical device safety and efficacy has been repeatedly raised. In each of the past several years, legislation regulating medical devices has been proposed and debated, and within recent months a report[5] by an internal government study group, which makes recommendations in this area, has been released. The high probability of device safety legislation being passed in the near future, coupled with the many years of experience in government regulation of pharmaceuticals, causes industry to be very concerned about this topic. That an appropriate type of government regulation and control is needed is not challenged by responsible industrial leaders. However, they fear procedures that will hinder innovation and create further delays to the already long period required to obtain adequate returns on investment.

The Task Group recognizes these valid concerns of industry. Because of the critical nature of many products in direct contact with the patient or used in the diagnosis of certain diseases, however, there must be greater assurance than that currently available that adequate protection is given to persons exposed directly or indirectly to biomedical products or services. This assurance must also extend to the doctor in the application and use of products because of his responsibility for the welfare of the patient.

> *There is an urgent need to develop a rational program that includes procedures for the development of standards and for the regulation and enforcement of medical device safety. The development must be a cooperative effort among government, professional societies, industry, and the medical and life scientists and must ensure that the initiative of private enterprise is not stifled.*

Specific Conclusions

Based on its study and experience, the Task Group developed the following specific conclusions to which its recommendations are addressed. For completeness, the enumeration below includes concise statements of the general conclusions discussed earlier.

1. Industry will respond quickly and effectively to develop, produce, and deploy biomedical engineering products and services when a reasonable profit can be forecast.
2. The present status of industrial activity in the biomedical engineering field is considerably below that which industrial capability can provide; the technology currently extant does not, in general, reflect the present state-of-the-art.
3. The foremost impediment to the expansion of industrial involvement in biomedical engineering is a lack of economic incentives brought about by the unique characteristics of the market for products and services.

4. Industry is not engaged in biomedical engineering research to any significant degree, leaving that realm of activity primarily to university and government laboratories.
5. Industry presently is not sufficiently involved in the formulation of biomedical engineering needs and potential solutions; it is unaware of priorities in the needs for development, and there is inadequate feedback of medical and technical problems and capability that evolve from medical needs, advancing technology, and industrial resources.
6. The programs of the various government agencies involved in the research and development of biomedical products and services have not provided the necessary amount of encouragement for industry-sponsored research and development.
7. Industry is confused about the differing patent policies of the various government agencies, does not appreciate the flexibility inherent in current policies, and is reluctant to utilize government-funded research and development until greater assurance of protection of industrial investments is obtained.
8. There are certain products and services that require direct government development subsidy or government-industry development cost sharing, yet mechanisms to provide this type of funding have not been adequately implemented.
9. Standards for and acceptance of uniform clinical evaluation procedures required for successful development and marketing of biomedical products have not been achieved.
10. A lack of knowledge and appreciation by each profession of the contributions that the other can make in this interdisciplinary endeavor is a major problem in the medical and engineering professions.
11. There exists a paucity of educational programs and access to relevant information that would create a common understanding between the professions; this paucity exists for all levels of activity, professional and managerial, as well as at the supportive level of the nurse and technician.
12. Qualified engineers have lacked the opportunities to work and accept engineering responsibilities in the medical and health care field; the need for professional engineering competence in these environments appears to be unrecognized.
13. The biomedical engineer has not yet adequately demonstrated or been given sufficient opportunity to demonstrate his ability to contribute within the industrial sector (i.e., in industrial employ).
14. While NIH support of PhD biomedical engineering training programs has been directed toward the national need for competent research-oriented

personnel, there is a lack of a sufficient number of competent product- and design-oriented biomedical engineers, trained at the BS and MS levels, who can function effectively in the industrial setting.

15. There is inadequate interaction between government agencies and the biomedical engineering industry, resulting in each having a lack of appreciation of the responsibilities, problems, and programs of the other.
16. The capability of and need for engineers to serve in responsible leadership positions in government biomedical research and development programs have not been fully recognized within government agencies.
17. Hospital and clinical personnel are inadequately trained in the use, operation, and maintenance requirements of technological products and services, and administrators do not appreciate the existence or impact of this inadequacy.
18. There are inadequate voluntary and regulated standards for the performance and safety of biomedical products and services, and effective enforcement procedures are yet to be established.

Recommendations

If the resources of United States' industry are to be effectively applied toward the achievement of adequate health care for all Americans, the medical and engineering professions, universities, industry, and government agencies must forge communication and action links to supplement the present patterns of interaction. The resulting national programs must come about through careful, deliberate, and collaborative design by all of the major participants. The development of this field should not be left to the uncoordinated or uninformed actions of isolated segments of society in the manner that currently predominates.

The Task Group on Industrial Activity offers the following objectives as national guidelines that should underlie the thrusts and directions taken by the collaborative bodies as they enhance the nation's biomedical engineering endeavor.

National Guidelines to Develop Industrial Biomedical Engineering Efforts

1. Provide for the dissemination of information on and a continuing assessment of medical and technical needs, priorities, state-of-the-art, product markets, resources, and capabilities on a national basis.
2. Bring about a national realization and recognition of the contributions the engineering profession has provided and can provide to the health care field.

3. Involve industry in the formulation of biomedical engineering problems admitting to technological solutions, stimulate means of attacking important problems not currently being pursued, foster adequate protection of industrial investments in research and development, and provide for uniform clinical evaluation procedures and facilities.
4. Clarify the respective roles of relevant government agencies and the capabilities and limitations of industry in the biomedical engineering field.
5. Ensure adequate standards and safety measures for biomedical products and services.

Formation of a National Overview Body in Biomedical Engineering

It is the strong belief of the Task Group that the coordination of the nation's biomedical engineering effort could best be achieved by the formation of a national advisory body, which would serve to create the focus and direction now lacking. *It is therefore the highest recommendation of the Task Group on Industrial Activity that an overview body, perhaps known as the National Biomedical Engineering Evaluation Panel, be immediately established.*

This Panel should be organized at the national level through some continuing and nonpolitical organization and supported through appropriate government agencies (e.g., NIH). It should be charged with the responsibility to:

1. Investigate national needs in the field of biomedical engineering as they relate to state-of-the-art knowledge and industrial technological capability.
2. Determine, as accurately as possible, the nature, market potential, and financial feasibility of proposed solutions to major needs in biomedical engineering.
3. Determine and recommend priorities and suggested time tables for the satisfaction of the major needs.
4. Recommend financial mechanisms best suited to the solutions of the national problems.
5. Identify more precisely the appropriate role of engineering in the life sciences, and investigate the means of providing engineers specifically for this role.

Membership should be drawn from the list of industrial, medical, engineering, and social experts most knowledgeable in the areas pertinent to the charge of responsibility. It would be expected that this national overview group would provide pertinent information and recommendations to government, industry, and the engineering-medical community on the current status of problems and solutions relating to research, development, and marketing of biomedical products; act as a liaison body to bring together interested parties who have a mutual interest in resolving critical problems; and avail itself to implement further studies, as warranted, to develop more specific information relevant to national efforts in biomedical engineering.

The study of biomedical engineering activity reported herein focused on industry, its perceptions and its problems. The problems that evolved need to be attacked immediately, and the Panel should undertake that challenge. Further, the Task Group urges that a parallel study of biomedical engineering as viewed by the medical profession and the health care delivery sector also be undertaken to completely delineate the national biomedical engineering endeavor. Only through such a study can an accurate evaluation be made of the priorities in a national effort that would mutually benefit both industry and the engineering-medical community. The Panel would be the appropriate body to undertake such a study.

Expansion and Delineation of the Roles of the National Institutes of Health and Other Government Agencies in Biomedical Engineering Activities

The historic role of NIH has been the support of fundamental research, concentrated on medical and biological approaches with a strong emphasis on such areas as physiology, biochemistry, and molecular biology. While biomedical engineering is an important component of a few programs within the Institutes (e.g., artificial heart program and certain activities within the National Institute of General Medical Sciences), the Task Group believes that NIH could immediately serve to enhance the national biomedical engineering endeavor by involving engineering more fully in the totality of NIH programs.* *The benefits to be derived by a more balanced involvement of engineers with medical scientists in biomedical research would contribute to the definition, identification, concept development, and applied research of devices and processes that could be of direct benefit to the health of the nation.*

However, NIH alone cannot optimally harness the full potential of our national biomedical engineering capability. Other federal agencies must also cooperate and assume their share of the task. The other offices within HEW, including the various divisions within the Health Services and Mental Health Administration and the Social and Rehabilitation Service, the Veterans' Administration, the Department of Defense, the National Aeronautics and Space Administration, the Atomic Energy Commission, the Department of Commerce, the National Science Foundation, the Office of Economic Opportunity, and the Office of Science and Technology each have responsibilities and resources to contribute.

It is the Task Group's belief, in fact, that there is currently no government agency that is, in a substantial way, attempting to close the gap between the

* Specific recommendations toward this end are found in Table 1.

identification of needs and the utilization of our national industrial capability to supply those needs. The National Institutes of Health has, as noted above, performed an outstanding job in fathering and directing the nation's medical research efforts. The institutes have not, in general, assumed the role of developers or deployers of biomedical engineering products and services, nor, perhaps, should they.

The Task Group, therefore, recommends that there should exist a government agency with a primary responsibility to develop and stimulate the deployment of biomedical engineering technology.

This agency would provide a focus within the government and would utilize and extend the efforts of other agencies (i.e., those noted above) that have implicit or explicit biomedical engineering programs. The agency should:

1. Have the necessary funding to carry prototypes through to the stage of multiple unit production and the means to conduct evaluations in clinical and research centers.
2. Involve industry at the earliest stages of prototype design, be able to issue exclusive, finite term licenses for production when required, and be able to subsidize needed production when the market potential is too limited for profits to be realized.
3. Encourage the utilization of industrial organizations to provide systems engineering, logistical studies, failure analyses, patient and material flow studies, and so on for the health care delivery system.

One existing agency that might be considered for this responsibility is the National Center for Health Services Research and Development (NCHSRD) whose statement[6] of origin and purpose would seem to admit to the acceptance of this effort. For reasons not explored by the Task Group, NCHSRD is not, however, currently fulfilling this role in a substantial or conclusive way, as evidenced by the fact that a significant number of the companies interviewed in the course of this study were unaware of the Center or its programs.

Tabular Summary of Specific Recommendations

The Task Group has provided two major and encompassing recommendations: the formation of a national advisory body and the establishment of a primary government agency responsible for the area under consideration. In addition and as an adjunct to these recommendations, the Task Group has formulated 18 specific proposals. They are variously addressed to nih, other government agencies, and the private sector composed of industry, universities, the health care delivery system, and the medical and engineering professions. These specific recommendations are listed in Table 1. Most of them are directly responsive to the inadequacies noted in the summary discussion and conclusions in the preceding chapter. All have their basis in the findings and analysis of the Task Group presented in Part III of this report.

Table 1 Specific Recommendations

Recommendation	Implemented by		
	NIH	Other Government Agencies	Private Sector
1. Encourage and contribute to the establishment and support of the National Biomedical Engineering Evaluation Panel	X	X	X
2. While continuing to fulfill its primary responsibility in basic research, NIH should broaden and make widely known its interest and responsibilities in the development of biomedical engineering products and services. A greater effort toward goal-oriented research would be consistent with this objective.	X		
3. Expand government in-house engineering competence by augmenting the biomedical engineering staff in the intramural and extramural programs of each institute and agency.	X	X	
4. Require realistic engineering involvement in government grants and contracts. Allowing engineers a greater opportunity to serve as principal investigators (in lieu of medical researchers) would be consistent with this objective.	X	X	
5. In addition to maintaining the current PhD training programs, support university trainee programs for design- and product-oriented biomedical engineers at the BS and MS levels.	X	X	X
6. Provide for engineering internships at NIH and other medical centers (both government and civilian) for practicing engineers from industry and for participants in the biomedical engineering programs of universities.	X	X	X
7. Provide for internships in industry to better identify the value and deficiencies of biomedical engineers in industrial situations.	X	X	X
8. Define and make widely known the responsibilities of each government agency in the research, development, evaluation, and deployment of biomedical engineering products and services.	X	X	
9. Encourage the developmental phase of high- priority biomedical engineering products by industry.	X	X	

(*Continued*)

Table 1 (Continued)

Recommendation	Implemented by		
	NIH	Other Government Agencies	Private Sector
10. Promote greater university-industry interaction in the development of biomedical products, the utilization of basic research, and the training of biomedical engineers.			X
11. Provide means for clinical evaluation to promote market acceptance of biomedical engineering products and services.	X	X	X
12. Unify the application of patent policies to profit as well as nonprofit organizations, and develop procedures (e.g., exclusive licenses) to encourage private risk capital investments in product development and deployment.	X	X	
13. Provide for an organization similar to the FDA to develop and regulate, along with industry, standards and safety measures for biomedical products and services.	X	X	X
14. Develop closer relationships between industry and the medical profession during the specification of product needs by expanded use of medical consultants in industry and collaborative industry-hospital and industry-clinic programs.			X
15. Create greater employment opportunities within industry to permit the demonstration by competent biomedical engineers of the contributions that they can provide.			X
16. Recognize the uniqueness of the biomedical engineering marketplace and develop the specific managerial techniques and personnel required to operate effectively within it.			X
17. Develop means to properly train the users of biomedical engineering instruments and provide for adequate maintenance, calibration, and repair services.		X	X
18. Provide support and designate a body (e.g., the National Biomedical Engineering Evaluation Panel) to conduct a study of the attitudes and position of the medical community toward engineering comparable to the study made of industry by this Task Group.	X	X	

In conclusion, the Task Group feels it imperative to note that the identification of many of the problems discussed above and recommendations for attacking them are not new. The referenced study of Jones[1] describes the problems of the marketplace and constraints that limit the development and deployment of potentially useful medical devices. The Panel on the Impact of New Technologies of the National Advisory Commission on Health Manpower gave a concise report in 1967,[7] which expressed the unfulfilled potential of technology and American industrial resources in health care. It describes the communication barrier, the need for adjustment of federal patent and licensing policies, the inadequate attention to product development, the nature of the individualistic market and of product evaluation and acceptance, and the need for greater effort in standards development. The Panel concludes by recommending the formation of two or more national bioengineering laboratories to provide the now-lacking national focus, coordination, and direction. Finally, an NIH report[8] published over a year ago states in part:

> *It is also the field of engineering, specifically biomedical engineering, which contributes enormously to the innovation of medical research and service. The potential contributions of engineering to medical research and service are exciting and far-reaching... meeting future manpower requirements calls for expansion of NIH programs in traditional fields; rapid expansion in newly emerging fields, such as biomedical engineering, . . . and expanded support of interdisciplinary study and research.*

Independently and collaboratively, many have studied and analyzed the problem. Their conclusions and recommendations are remarkably similar. If the federal government and its constituent agencies, such as the National Institutes of Health, are willing to assume the challenge, the road maps have been charted. In the view of the Task Group on Industrial Activity, the time for concerted action is now.

Source: *National Academy of Engineering. (1971). An* Assessment of industrial activity in biomedical engineering. *Reproduced with permission from the National Academy of Sciences. Courtesy of the National Academies Press, Washington, D.C.*

Questions to Consider

This document presents some of the earliest work on the emerging field blending health care, medicine, and engineering. The study documented a 1969 effort to assess the biomedical engineering landscape, work that predated federal legislation regarding medical devices.

Points to reflect/discuss:

- Although not reprinted here, the Preface of this article states, "...often heard that the communication barrier that exists between the technologist and the physician hinders the collaboration between the two." This statement was written about 50 years ago. Have things changed? Might this observation be part of human nature and insurmountable?
- In the 50 years since this report published, has an "atmosphere of mutual respect and acceptance" been established? What evidence could be provided to support or deny this expectation?
- The limited use of medical technology at the time of publication is evident in shaping the comment, "certain devices need to be produced in small quantities." At the time, only the sickest patients in an intensive care unit might have their cardiac activity monitored. Intravenous treatments were provided by gravity, not a mechanical pump. The extraordinary expansion of medical devices utilized at the bedside and as part of diagnosis and treatment cannot be imagined in a present day suffused with technology in almost every aspect of daily living. How does the environment of limited medical device utilization of that era shape the findings of the study? If the authors could see into present day and understand the impact of technology in health care, might the recommendations be different?
- The authors extended the concerns expressed about limited technology need/utilization with the italicized section regarding a "viable marketplace." Has a viable marketplace for medical technology been established? Most would agree, but when did this happen? When did the shift from human-centric care (e.g., gravity-fed intravenous delivery regularly monitored by nurses) to technology-mediated care occur?
- The report mentions medical device safety. How does this tie into the 1971 *Ladies' Home Journal* article release, the founding of AAMI, and the development of standards by the NFPA?
- The report describes an impediment to biomedical engineering as an industry caused by the lack of "economic incentives" (conclusion point 3). Has this changed? How is this change evident?
- Review the study recommendations, and identify which recommendations were eventually achieved (with or without direct intervention). Note recommendation number 17 regarding the equipment support. Has this been achieved almost 50 years later? Reflect on the number of academic programs that focus on the safe and effective use of medical devices, at either the engineering or technician level? Are the number of programs adequate for the amount of hospitals in the United States? Across the globe?
- Although not included in this reprint, the original report describes three broad areas of engineering as applied to medicine and biology. The language utilized in this report appendix has been cited by some clinical engineering professionals as the earliest characterization of clinical engineering.

  "1. The application of engineering concepts and technology to the scientific inquiry into biological phenomena as a basis for advancing the understanding of biological systems and medical practices."—research into new discovery and new ways of knowing

  "2. The utilization of engineering concepts and technology in the development of instrumentation, materials, diagnostic and therapeutic devices, artificial organs, and other constructs relevant to applications in biology and medicine." —device innovation

"3. The application of engineering concepts, methodology, and technology to the improvement of health service delivery systems in the broad context of interrelated institutions (i.e., hospitals, clinics, governmental units, universities, industry) as well as within the specific confines of individual components of the health care system."—clinical applications of engineering principles = clinical engineering

Are these broad categories still appropriate? Are these accurate? It is important to note that the term *clinical engineering* is not located in this report. Might the use of the term have advanced the profession?

Article 2: Human resources for medical devices, the role of biomedical engineers. Geneva: World Health Organization (2017)

Preface

Health technologies are essential for a fully functioning health system. Medical devices, in particular, are crucial in the prevention, diagnosis, treatment and palliative care of illness and disease, as well as patient rehabilitation. Recognizing this important role of health technologies, the World Health Assembly adopted resolution WHA60.29 in May 2007. The resolution covers issues arising from the inappropriate deployment and use of health technologies, and the need to establish priorities in the selection and management of health technologies, specifically medical devices.

This publication has been produced in accordance with the request to the World Health Organization (WHO) by Member States in the WHA60.29.*(1)* The resolution:

1. URGES Member States:

> *"(1) to collect, verify, update and exchange Information on health technologies, In particular medical devices, as an aid to their prioritization of needs and allocation of resources;*
>
> *(2) to formulate as appropriate national strategies and plans for the establishment of systems for the assessment, planning, procurement and management of health technologies, in particular medical devices, in collaboration with personnel involved in health-technology assessment and biomedical engineering..."*

2. REQUESTS the Director General:

> *"to work with interested Member States and WHO collaborating centres on the development, in a transparent and evidence-based way, of guidelines and tools, including norms, standards and a standardized glossary of definitions relating to health technologies in particular medical devices"(2)*

By adopting this resolution, delegations from Member States acknowledged the importance of health technologies for achieving health-related development goals, urging expansion of expertise in the field of health technologies, in particular medical devices, and requesting that WHO take specific actions to support Member States.

One of WHO's strategic objectives is to "ensure improved access, quality and use of medical products and technologies." To meet this objective, WHO and partners have been working towards devising an agenda, an action plan, tools and guidelines to increase access to good quality, appropriate medical devices. The *Medical device technical series* developed by WHO already includes the following publications:

- Development of medical devices policies*(3)*
- Health technology assessment for medical devices*(4)*
- Health technology management
  - Needs assessment process of medical devices*(5)*
  - Procurement process, resource guide*(6)*
  - Medical devices donations, considerations for solicitation and provision*(7)*
  - Introduction to medical equipment inventory management*(8)*
  - Medical equipment maintenance overview*(9)*
  - Computerized maintenance management systems*(10)*
- Priority medical devices
  - Interagency list of priority medical devices for essential interventions on reproductive, maternal, new born and child *health(11)*
  - Protective equipment for Ebola*(12)*

And will include, in 2017, the following publications, which are under development:

- Human resources for medical devices (the present document)
- Model regulatory framework for medical devices
- Priority medical devices for cancer management

These documents are intended for use by policy-makers at ministries of health, biomedical engineers, health managers, donors, nongovernmental organizations and academic institutions involved in health technology at the district, national, regional and global levels.

Medical device technical series – methodology

The first documents in the *Medical device technical series* were written by international experts in their respective fields, and reviewed by members of the Technical Advisory Group on Health Technology (TAGHT). The TAGHT was established in 2009 to provide a forum for both experienced professionals and country representatives to develop and implement the appropriate tools and documents to meet the objectives of the Global Initiative on Health

Technologies (GIHT). The group met on three occasions. The first meeting was held in Geneva in April 2009 to prioritize which tools and topics most required updating or developing. A second meeting was held in Rio de Janeiro in November 2009 to share progress on the health technology management tools under development since April 2009, to review the current challenges and strategies facing the pilot countries, and to hold an interactive session for the group to present proposals for new tools, based on information gathered from the earlier presentations and discussions.

The last meeting was held in Cairo in June 2010 to finalize the documents and to help countries develop action plans for their implementation. In addition to these meetings, experts and advisers have collaborated through an online community to provide feedback on the development of the documents. The concepts addressed in the *Medical device technical series* were discussed further during the First Global Forum on Medical Devices in September 2010. Stakeholders from 106 countries made recommendations on how to implement the information covered in this series of documents at the country level.*(13)* These extensive discussions formed the background for the current book in the series: *Human resources for medical devices*.

For the development of the current book the following activities took place:

1. Global BME surveys, the outcomes of which are highlighted in this book and results presented in the WHO medical devices website and in the WHO Global Health Observatory.*(14)*

In 2009, WHO launched "Biomedical engineering global resources,"(15) a WHO programme to gather information on academic programmes, professional societies and the status of BME worldwide.

Since the programme began, WHO has collected and compiled data in five stages with different institutions, colleagues and tools, all coordinated by Adriana Velazquez from WHO. Table 1 outlines the details for each stage.

To collect the information, an electronic survey was tailored for each of the four stages of the project. Surveys were distributed through a network of over 3200 BME professionals using the WHO medical devices Listserv tool.*(16)* Survey participants were encouraged to resend the survey to other key informants with the intention of collecting a greater amount of current global information.

The surveys were consistently structured into three sections: (1) country profile; (2) educational institutions; and (3) professional societies. An additional section (4) was added to the 2015 version targeting international organizations and including questions regarding female presence within the profession. The survey was designed so respondents only completed the section for which they held information, ensuring that information was supplied by

Table 1 WHO surveys on biomedical engineering availability

Year	Lead	Methodology	Outcome
2009	S Calil (University of Campinas)	Conducted active search for universities and professional societies	First global database of BME resources (http://who.ceb.unicamp.br/)
2013	Survey developed by D Desai (Boston University) with support from J Barragan (WHO), S Mullally (WHO) and N Jimenez (WHO)	Developed survey with IT tool that was sent to WHO BME contacts	Updated global database of BME resources (http://apps.who.int/medical_devices/edu/)
2014	C Long (WHO) and R Magjarevic (IFMBE)	Developed survey with LimeSurvey tool to collect the specific number of biomedical engineers per country and integrate it with information from the IFMBE database of national societies	Information on biomedical engineer density (http://hqsudevlin.who.int:8086/data/node.main.HRMDBIO?lang = en)
2015	D Rodriguez (WHO), M Smith (WHO) and R Martinez (WHO)	Developed survey with WHO DataForm tool. Collected data were integrated with data from the 2009, 2013 and 2014 investigations	Organization of data from 85 countries
2016	D Rodriguez (WHO), C Soyars (WHO) and R Martinez (WHO)	Compiled information from all stages of data collection	Annex 1,2 and 3 of this publication

national health authorities or similar for section 1, BME academics for section 2, professional societies secretariats for section 3 and international organization focal points for section 4. For the data analysis, 140 fully completed survey forms were retrieved from the data collection tool (WHO DataForm). *(17)* Additional information was collected from IFMBE reports, other BME professional organizations and web-based inquiries.

The collected data are available at the WHO medical devices site: http:// www.who.int/ medical_devices/support/en/ and the density of biomedical engineers is available on WHO Global Health Observatory (GHO) as a new indicator of global health: http://apps.who.int/gho/data/node.imr.HRH_40?lang = en and further information can be found at: http://apps.who.int/gho/data/node.main.504?lang = en.

WHO intends to collect data annually and expects to retrieve data from unreported countries, validate reported data, maintain accurate estimates and record global trends in BME.

2. Professionals of different branches of biomedical engineering (BME) were invited by WHO in 2013 to become chapter coordinators. The chapter coordinators wrote sections and organized the input of 40 collaborators from around the world who contributed their expertise on design, development, regulation, assessment, management and safe use of medical devices.
3. The Second Global Forum on Medical Devices, held in November 2013, included a session focused specifically on human resources for medical devices and the required survey methodology addressing the role of the biomedical engineer.*(18)* The forum had diverse country representation, with a total of 572 participants representing 103 countries.
4. Formal discussions about the role of BME in Hangzhou, China, at a global BME summit on 23 October 2015, following the First International Clinical Engineering and Health Technology Management Congress, which was sponsored by the Chinese Society of Biomedical Engineering. The congress was organized by a leadership panel of 25 representatives of national and international organizations related to BME. The group produced a document that lays out the scope of professional activities of biomedical engineers and provides a rationale and evidence for the recognition and proper classification of *BME.(19)*
5. Review of whole text by international experts on biomedical engineering.

Scope

WHO invited global collaboration of biomedical engineering professionals to support the development of the present publication in order to describe the different roles of biomedical engineers as specialized human resources on medical devices, in order to support their classification in Member States as well as labour and other organizations.

It should be noted that there are many professionals who use medical devices, for example radiographers, sonographers, laboratory technicians, surgeons, anaesthetists, ophthalmologists, orthopaedists, neurologists, radiologists, radiotherapists, pathologists, nurses, and almost all health-care workers. And, there are others who manage their supply, such as procurement and logistics officers, others that evaluate them as health economists, others that support the design and manufacture, including engineers, industrial designers, chemists, physicists, etc.

This publication addresses only the role of the biomedical engineer in the development, regulation, management, training and use of medical devices in general, and it should be noted that this happens with interdisciplinary collaboration according to rules and regulations in every country.

Definitions

Recognizing that there are multiple interpretations for the terms listed below, they are defined as follows for the purposes of this publication of the WHO *Medical device technical series.*

"Biomedical engineering" includes equivalent or similar disciplines, whose names might be different, such as medical engineering, electromedicine, bioengineering, medical and biological engineering and clinical engineering.

Health technology: The application of organized knowledge and skills in the form of devices, medicines, vaccines, procedures and systems developed to solve a health problem and improve quality of life.*(20)* It is used interchangeably with health-care technology.

Medical device: An article, instrument, apparatus or machine that is used in the prevention, diagnosis or treatment of illness or disease, or for detecting, measuring, restoring, correcting or modifying the structure or function of the body for some health purpose. Typically, the purpose of a medical device is not achieved by pharmacological, immunological or metabolic means.*(21)* This category includes medical equipment, implantables, single use devices, in vitro diagnostics and some assistive technologies.

Medical equipment: Used for the specific purposes of diagnosis and treatment of disease or rehabilitation following disease or injury; it can be used either alone or in combination with any accessory, consumable or other medical equipment. Medical equipment excludes implantable, disposable or single-use medical devices. Medical equipment is a capital asset and usually requires professional installation, calibration, maintenance, user training and decommissioning, which are activities usually managed by clinical engineers.*(22)*

Health technology assessment (HTA): The term refers to the systematic evaluation of properties, effects and/or impacts of health technology. It is a multidisciplinary process to evaluate the social, economic, organizational and ethical issues of a health intervention or health technology. The main purpose of conducting an assessment is to inform a policy decision-making process.*(23)*

Biomedical engineering (BME): Medical and BME integrates physical, mathematical and life sciences with engineering principles for the study of biology, medicine and health systems and for the application of technology to improve health and quality of life. It creates knowledge from molecular to organ systems levels, develops materials, devices, systems, information approaches, technology management and methods for assessment and evaluation of technology, for the prevention, diagnosis and treatment of disease, for health-care delivery and for patient care and rehabilitation.*(24)*

Among the sub-specialities of biomedical engineering, or engineering and technology related professions, are the following:

Clinical engineer: In some countries, this defines the biomedical engineer that works in clinical settings. The American College of Clinical Engineering defines a clinical engineer as, "a professional who supports and advances patient care by applying engineering and managerial skills to health care technology".*(25)* The Association for the Advancement of Medical Instrumentation describes a clinical engineer as, "a professional who brings to health-care facilities a level of education, experience, and accomplishment which will enable him to responsibly, effectively, and safely manage and interface with medical devices, instruments, and systems and the user thereof during patient care...".*(26)*

Biomedical engineering technician/technologist (BMET): Front-line practitioners dedicated to the daily maintenance and repair of medical equipment in hospitals, meeting a specified minimum level of expertise. BMETs who work exclusively on complex laboratory and radiological equipment may become certified in their specialism, without needing to meet the more general professional engineering requirements. The difference between a technician and a technologist relates to the level and number of years of training. Normally technicians train for two years, technologists for three years, but this can differ per country.

Rehabilitation engineers: Those who design, develop and apply assistive devices and technologies are those whose primary purpose is to maintain or improve an individual's functioning and independence to facilitate participation and to enhance overall well-being.*(27)*

Biomechanical engineers: Biomechanical engineers apply engineering principles to further the understanding of the structure of the human body, the skeleton and surrounding muscles, the function and engineering properties of the organs of the body, and use the knowledge gained to develop and apply technologies such as implantable prostheses and artificial organs to aid in the treatment of the injured or diseased patient to allow them to enjoy a better quality of life.

Bioinstrumentation engineers: Bioinstrumentation engineers specialize in the detection, collection, processing and measurement of many physiological parameters of the human body, from simpler parameters like e.g. temperature measurement and heart rate measurement to the more complex such as quantification of cardiac output from the heart, detection of the depth of anaesthesia in the unconscious patient and neural activity within the brain and central nervous system. They have been responsible for the development and introduction of modern imaging technologies such as ultrasound and magnetic resonance imaging (MRI).

In order to raise awareness on the importance of biomedical engineers within health systems worldwide, WHO is producing the present publication, in conjunction with a global BME survey. WHO, along with international professional organizations, such as IFMBE, is also advancing a proposal to update the ILO classification, *(28)* requesting a specific classification for biomedical engineers as a distinct professional category.

Acknowledgements

The publication *Human resources for medical devices: The role of biomedical engineers* was developed under the coordination of Adriana Velazquez Berumen, WHO Senior Advisor and Focal Point on Medical Devices, with the supervision of Gilles Forte and Suzanne Hill in the WHO Essential Medicines and Health Products Department, and from James Campbell and Giorgio Cometto, in the Health Workforce Department under the Health Systems and Innovation Cluster of the World Health Organization.

Each chapter of the book was authored by an expert in the field, along with contributions from collaborators from different regions of the world and diverse sectors of the BME field during 2014 and 2015.

The BME 2015 survey was developed in January 2015 by Daniela Rodriguez-Rodriguez and Megan Smith. The first results were presented in the World Congress on Biomedical Engineering and Medical Physics in Toronto, June 2015, and are included in this publication.

The first draft of this book was edited in 2015, for discussion at the First International Clinical Engineering and Health Technology Management Congress in China.*(29)*

The International Federation of Medical and Biological Engineering (IFMBE, an NGO in official relations with WHO, chaired by James Goh, Singapore) generously supported the development of this publication, including technical editing and graphic design, and appointed Fred Hosea to support the compilation of the chapters, conducting teleconferences with the authors and collaborators.

Data analytics and editorial reviews were done by Daniela Rodriguez-Rodriguez, Ileana Freige and Ricardo Martinez. Final integration of edits was done by Anna Worm and Adriana Velazquez.

WHO is grateful to the following chapter coordinators for their contributions of content, editorial advice, and for managing the input from contributors from around the world; and to the chapter contributors for their expert research and information submitted.

Biomedical engineers as human resources for health

Chapter coordinator: Adriana Velazquez

Chapter contributors: Michael Flood, Australia, Fred Hosea, Ecuador; Anna Worm, Benin.

1. **Global dimensions of biomedical engineering**
   Chapter coordinator: Adriana Velazquez Berumen, WHO.
   Chapter contributors: Michael Flood, Australia; Daniela Rodriguez-Rodriguez, Mexico; Ratko Magjarevic, Croatia; Megan Smith, Canada.
2. **Education and training**
   Chapter coordinator: Herbert F Voigt, United States of America.
   Chapter contributors: Fred Hosea, Ecuador; Kenneth I Nkuma-Udah, Nigeria; Nicolas Pallikarakis, Greece; Kang Ping Lin, China; Mario Secca, Mozambique; Martha Zequera, Colombia.
3. **Professional associations**
   Chapter coordinator: Yadin David, United States of America.
   Chapter contributors: Stefano Bergamasco, Italy; Joseph D Bronzino, United States of America; Saide Calil, Brazil; Steve Campbell, United States of America; Tom Judd, United States of America; Niranjan Khambete, India; Kenneth I Nkuma-Udah, Nigeria; Paolo Lago, Italy; Consuelo Varela Corona, Cuba; Herb Voigt, United States of America; Zheng Kun, China; Zhou Dan, China.
4. **Certification**
   Chapter coordinator: Mario Medvedec, Croatia.
   Chapter contributor: James O Wear, United States of America; Anthony Chan, Canada.
5. **Development of medical devices policy**
   Chapter coordinator: Adriana Velazquez Berumen, WHO.
6. **Medical device research and innovation**
   Chapter coordinator: Mladen Poluta, South Africa.
   Chapter contributors: David Kelso, United States of America; Einstein Albert Kesi, India; Victor Konde, South Africa; Paul LeBarre, Denmark; Amir Sabet Sarvestani, United States of America.

7. **Role of biomedical engineers in the regulation of medical devices**
   Chapter coordinator: Saleh Al Tayyar, Saudi Arabia.
   Chapter contributor: Michael Gropp, United States of America. Josephina Hansen, WHO.
8. **Role of biomedical engineers in the assessment of medical devices**
   Chapter coordinator: Reiner Banken, Canada.
   Chapter contributors: Karin Diaconu, Bulgaria; Erin Holmes, United States of America; Debjani Mueller, South Africa; Leandro Pecchia, Italy.
9. **Role of biomedical engineers in the management of medical devices**
   Chapter coordinators: Corrado Gemma, Italy; Paolo Lago, Italy.
   Chapter contributors: Firas Abu-Dalou, Jordan; Roberto Ayala, Mexico; Stefano Bergamasco, Italy; Valerio Di Virgilio, Panama; Tom Judd, United States of America; Niranjan Khambete, India; Sitwala Machobani, Zambia; Mario Medvedec, Croatia; Shauna Mullally, Canada; Valentino Mvanga, United Republic of Tanzania; Ebrima Nyassi, Gambia; Ledina Picari, Albania; Adriana Velazquez Berumen, WHO.
10. **Role of biomedical engineers in the evolution of health-care systems**
    Chapter coordinator: Elliot Sloane, United States of America.
    Chapter contributor: Fred Hosea, Ecuador.

Acronyms and abbreviations

AAMI	Association for the Advancement of Medical Instrumentation
ABEC	African Biomedical Engineering Consortium
ABET	Accreditation Board for Engineering and Technology
ACCE	American College of Clinical Engineering
ACPSEM	Australasian College of Physical Scientists and Engineers in Medicine
AEMB	Alpha Eta Mu Beta
AFR	African Region (WHO)
AFPTS	Association Francophone des Professionnels des Technologies de Sante
AHF	Asian Hospital Federation
AHT	assistive health technology
AIIC	Associazione italiana ingegneri clinici (Italy)
AIMBE	American Institute for Medical and Biological Engineering
AMR	Region of the Americas (WHO)
ANS	affiliated national society
ANVISA	Agencia Nacional de Vigilancia Sanitaria (Brazil)
BME	biomedical engineering
BMES	Biomedical Engineering Society
BMET	biomedical engineering technician

CAHTMA	Commission for the Advancement in Healthcare Technology Management in Asia
CBET	certified biomedical equipment technician
CCE	certified clinical engineer
CE	clinical engineer
CED	Clinical Engineering Division (IFMBE)
CEO	chief executive officer
CET	certified electronics technician/certified engineering technologist
CFO	chief financial officer
CIO	chief information officer
CIS	cardiology information system
CLES	clinical laboratory equipment specialist
ComHEEG	High-Level Commission on Health Employment and Economic Growth
CORAL	Consejo Regional de Ingenieria Biomedica para America Latina
CPD	continuing professional development
CQI	continuous quality improvement
CRES	certified radiological equipment specialist
CSE	clinical systems engineer
CTO	chief technology officer
DBE	Directorate of Biomedical Engineering (Jordan)
EAMBES	European Alliance for Medical and Biological Engineering and Science
ECA	Economic Commission for Africa
ECG	electrocardiogram
ECTS	European Credit Transfer System
EESC	European Economic and Social Committee
EHEA	European Higher Education Area
EHR	Electronic health records
EMBS	Engineering in Medicine and Biology Society
EMR	Eastern Mediterranean Region (WHO)
ENA	Eastern Neighbouring Area
ESEM	European Society of Engineering and Medicine
EUnetHTA	European Network for HTA
EUR	European Region (WHO)
EWH	Engineering World Health
FDA	Food and Drug Administration (USA)
GHO	Global Health Observatory (WHO)
GIHT	Global Initiative on Health Technology
GHTF	Global Health Task Force
GMP	good manufacturing practice

HIC	High income country
HIT	health information technology
HRM	health risk management
HTA	health technology assessment
HTAD	Healthcare Technology Assessment Division (IFMBE)
HTAi	Health Technology Assessment international
HTM	Health Technology Management
HTCC	Health Technology Certification Commission (ACCE)
HTTG	Health Technology Task Group (IUPESM)
IAIE	International Atomic Energy Agency
ICC	Certification Commission for Clinical Engineering and Biomedical Technology
ICMCC	International Council on Medical and Care Compunetics
ICSU	International Council of Science
IEEE	Institute of Electrical and Electronics Engineers
IFMBE	International Federation for Medical and Biological Engineering
ILO	International Labour Organization
IMDRF	International Medical Device Regulators Forum
INAHTA	International Network of Agencies for Health Technology Assessment
INCOSE	International Council on Systems Engineering
IOMP	International Organization for Medical Physicists
ISO	International Standards Organization
IRB	International Registration Board
ISCO	International Standard Classification of Occupations
ITIL	Information Technology Infrastructure Library
IUPESM	International Union for Physical and Engineering Sciences in Medicine
LIC	Low income country
LIS	laboratory information system
LMIC	low- and middle-income countries
MNCH	maternal, newborn and child health
MRA	mutual recognition agreement
MRI	magnetic resonance imaging
NEA	national examining authority
NGO	nongovernmental organization
OEM	original equipment manufacturer
PACS	picture archiving and communication system
PAHO	Pan American Health Organization
PET	positron emission tomography
PPE	personal protective equipment

PPM	planned preventive maintenance
QA	quality assurance
QMS	quality management systems
R&D	research and development
RAC	Regulatory Affairs Certification
RAPS	Regulatory Affairs Professionals Society
RCEC	Registered Clinical Engineer Certification
RedETSA	Red de Evaluacion de Tecnologias Sanitarias de las Americas
RIS	radiology information system
SDG	Sustainable Development Goals
SEAR	South-East Asia Region (WHO)
SoS	systems of systems
SoSE	systems of systems engineering
TAGHT	Technical Advisory Group on Health Technology (WHO)
THET	Tropical Health & Education Trust (UK)
TSBME	Taiwan Society for Biomedical Engineering
UCT	University of Cape Town
UHC	universal health coverage
UNECA	United Nations Economic Commission for Africa
UNFPA	United Nations Population Fund
UNICEF	United Nations Children's Fund
UNOPS	United Nations Office for Project Services
V&V	verification and validation
WHO	World Health Organization
WMDO	World Medical Device Organization
WPR	Western Pacific Region (WHO)

Country acronyms

AFG	Afghanistan
ALB	Albania
ARE	United Arab Emirates
ARG	Argentina
AUS	Australia
AUT	Austria
BEL	Belgium
BEN	Benin
BFA	Burkina Faso
BGD	Bangladesh
BGR	Bulgaria
BHR	Bahrain
BIH	Bosnia and Herzegovina

BLZ	Belize
BOL	Bolivia (Plurinational State of)
BRA	Brazil
BRB	Barbados
BRU	Brunei
BTN	Bhutan
CAN	Canada
CHE	Switzerland
CHL	Chile
CHN	China
CIV	Cote d'Ivoire
CMR	Cameroon
COD	Democratic Republic of the Congo
COL	Colombia
CUB	Cuba
CYP	Cyprus
CZE	Czech Republic
DEU	Germany
DJI	Djibouti
DNK	Denmark
DOM	Dominican Republic
DZA	Algeria
ECU	Ecuador
EGY	Egypt
ESP	Spain
EST	Estonia
ETH	Ethiopia
FIN	Finland
FRA	France
FSM	Micronesia (Federated States of)
GBR	United Kingdom
GEO	Georgia
GHA	Ghana
GIN	Guinea
GMB	Gambia
GRC	Greece
GRD	Grenada
GTM	Guatemala
GUY	Guyana
HND	Honduras
HRV	Croatia

HTI	Haiti
HUN	Hungary
IDN	Indonesia
IND	India
IRL	Ireland
ISL	Iceland
ISR	Israel
ITA	Italy
JAM	Jamaica
JOR	Jordan
JPN	Japan
KEN	Kenya
KGZ	Kyrgyzstan
KIR	Kiribati
KOR	Republic of Korea
LAO	Lao People's Democratic Republic
LBN	Lebanon
LBR	Liberia
LKA	Sri Lanka
LTU	Lithuania
LVA	Latvia
MDA	Republic of Moldova
MEX	Mexico
MKD	The former Yugoslav Republic of Macedonia
MNE	Montenegro
MNG	Mongolia
MOZ	Mozambique
MYS	Malaysia
NAM	Namibia
NGA	Nigeria
NLD	Netherlands
NOR	Norway
NPL	Nepal
NZL	New Zealand
PAK	Pakistan
PAN	Panama
PER	Peru
PHL	Philippines
POL	Poland
PRT	Portugal
PRY	Paraguay

ROU	Romania
RUS	Russian Federation
RWA	Rwanda
SAU	Saudi Arabia
SDN	Sudan
SEN	Senegal
SGP	Singapore
SLE	Sierra Leone
SLV	El Salvador
SRB	Serbia
SUR	Suriname
SVK	Slovakia
SVN	Slovenia
SWE	Sweden
SWZ	Swaziland
TCD	Chad
THA	Thailand
TLS	Timor-Leste
TTO	Trinidad and Tobago
TUN	Tunisia
TUR	Turkey
TZA	United Republic of Tanzania
UGA	Uganda
UKR	Ukraine
URY	Uruguay
USA	United States of America
VEN	Venezuela (Bolivarian Republic of)
VNM	Viet Nam
VUT	Vanuatu
YEM	Yemen
ZAF	South Africa
ZMB	Zambia

Executive summary

Continuous developments in science and technology are increasing the availability of thousands of medical devices - all of which should be of good quality and used appropriately to address global health challenges. It is recognized that medical devices are becoming ever more indispensable in health-care provision and among the key specialists responsible for their design, development, regulation, evaluation and training in their use – are biomedical engineers.

In this book, part of the *Medical device technical series*, WHO presents the different roles the biomedical engineer can have in the life cycle of a medical device, from conception to use.

It is important to mention that for this publication, the concept "biomedical engineer" includes medical engineers, clinical engineers and related fields as categorized in different countries across the world and encompasses both university level training as well as that of technicians.

Working together with other health-care workers, biomedical engineers are part of the health workforce supporting the attainment of the Sustainable Development Goals, especially universal health coverage.

This book has two parts. The first looks at the biomedical engineering profession globally as part of the health workforce: global numbers and statistics, and professional classification, general education and training, professional associations and the certification process.

The second part addresses all the different roles that the biomedical engineer can have in the life cycle of the technology, from research and development, and innovation, mainly undertaken in academia; the regulation of devices entering the market; the assessment or evaluation in selecting and prioritizing medical devices (usually at national level); to the role they play in the management of devices from selection and procurement, to safe use in health-care facilities.

Finally, the annexes present comprehensive information on academic programmes, professional societies and relevant WHO and UN documents related to human resources for health, as well as the reclassification proposal for ILO.

This publication can be used to encourage the availability, recognition and increased participation of biomedical engineers as part of the health workforce, particularly following the recent adoption of the recommendations of the UN High-Level Commission on Health Employment and Economic Growth, the WHO Global Strategy on Human Resources for Health, and the establishment of national health workforce accounts. The document also supports the aim of reclassification of the role of the biomedical engineer as a specific engineer that supports the development, access and use of medical devices, within the national, regional and global occupation classification system.

The biomedical engineer can play a crucial role in supporting the best and most appropriate use of medical technologies to help in achieving universal health coverage and the targets of the Sustainable Development Goals. Biomedical engineers can take their share of responsibility and develop continuously better competencies to help achieve these goals, so vital for those in most need in and with least resources.

Biomedical engineers as human resources for health

Biomedical engineering is one of the more recently recognized disciplines in the practice of engineering. It is a field of practice which brings many, if not all of the classical fields of engineering together to assist in developing a better understanding of the physiology and structures of the human body, and to support the knowledge of clinical professionals in prevention, diagnosis and treatment of disease and modifying or supplementing the anatomy of the body with new devices and clinical services.

Biomedical engineering is considered as the profession responsible for innovation, research and development, design, selection, management and safe use of all types of medical devices, including single-use and reusable medical equipment, prosthetics, implantable devices and bionics, among others.

A key objective of biomedical engineers is to have devices that are of good quality, effective for the intended purpose, available, accessible and affordable. When these objectives are met and devices are used safely, patients' lives may be saved, quality of life increased and there will be positive economic outcomes; the final goal is attainment of better levels of care. The prerequisites for this to happen are health technology policies in national health plans, available human and financial resources, and scientific and technological advances that lead to usable knowledge and information. The interrelations of these concepts are presented in Figure 1.

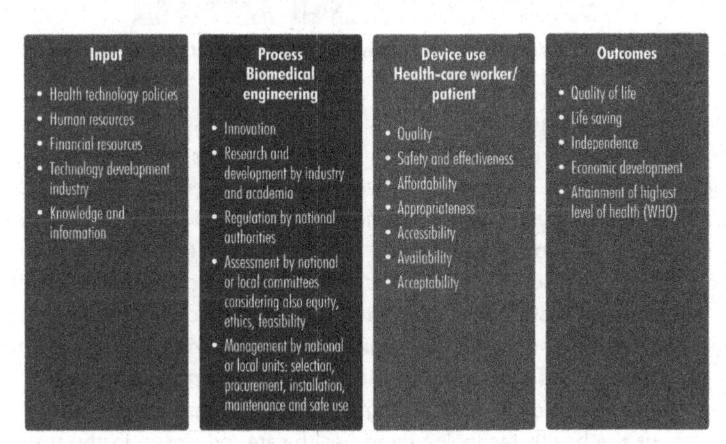

Figure 1 Medical devices process from policies to health outcomes

The practice of BME is not new. Indeed, the first known use of a functioning medical prosthesis for a toe is traceable to the African continent; specifically, to Egypt. It could be argued that Leonardo da Vinci (1442–1519) and many of the ancient philosophers could be considered among the first biomedical engineers. Among his other interests, da Vinci studied the anatomy of the human skeleton, the muscles and sinews of the body.

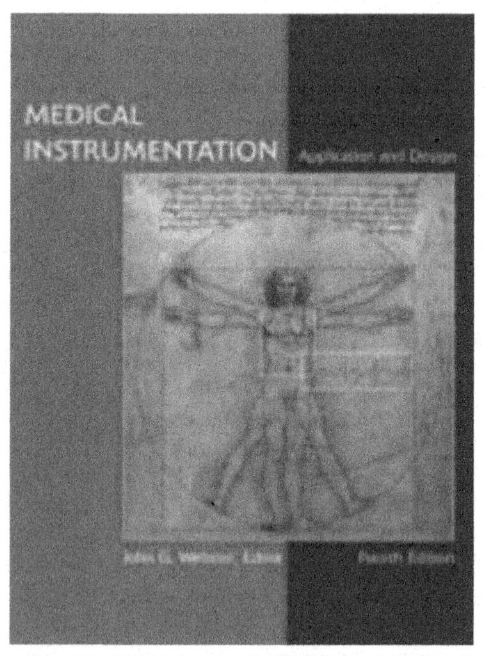

Figure 2 Da Vinci's ***Vitruvian Man*** drawing (from around 1490) featured on a key reference book on medical instrumentation

Responsibilities and roles

Biomedical engineering professionals are key players in developing and advancing the usage of medical devices and clinical services. Depending on their training and sector of employment, the responsibilities of biomedical engineering professionals can include overseeing the research and development, design, safety and effectiveness of medical devices/systems; selection and procurement, installation, integration with electronic medical records systems, daily operations monitoring, managing maintenance and repairs, training for safe use and upgrading of medical devices available to health-care stakeholders. Biomedical engineering professionals are employed widely throughout the health technology and health-care industries, in the research and development (R&D) of new technologies, devices and treatment modalities, in delivery of health-care in hospitals and other institutions, in academia, government institutions and in national regulatory agencies.

Research and development

When employed in research and development including both industry as academic institutions, the role of the biomedical engineering professionals is typically one of bringing together the specialist skills of the other engineering disciplines such as mechanical, materials, signal processing and others, using their broad engineering knowledge, coupled with their knowledge of medical practice, the human physiology and body structures to ensure the end result of their collective work is a product that is safe, effective and performs as intended for the benefit of the patient. As devices become "smarter" through the inclusion of increasingly powerful hardware and software capabilities, devices can take on increasingly comprehensive monitoring, alert and control functions that define clinical best practices. This "smart device" revolution is extending the domain of BME into wider and wider realms of creativity and professional practice, extending health-care services far beyond the hospital.

Health-care providers

When biomedical engineering professionals are employed in health-care institutions, their roles can include asset management, equipment selection, installation and maintenance, planning of clinical areas for health-care delivery, support other healthcare professionals to define appropriate technologies for patient diagnostic, treatment and rehabilitation as well as development of specialized instruments or devices for research or treatment and customized, patient-specific devices.

Government

Many biomedical engineering professionals are also engaged by government such as ministries of health, working on central or regional level health-care technology management, or governmental organizations such as health technology assessment or regulatory agencies, where their skills are applied to the evaluation for selection of public procurement, reimbursement schemes or examination or testing of medical devices to ensure those to be placed on the market are safe and in compliance with international standards and regulatory requirements.

Industry

A part of the biomedical engineering professionals in industry work in R&D. Another branch of activities is in sales and service, where biomedical engineering professionals can play a role in assuring customers are supported in making choices and providing after-sales service, like training, maintenance and repair.

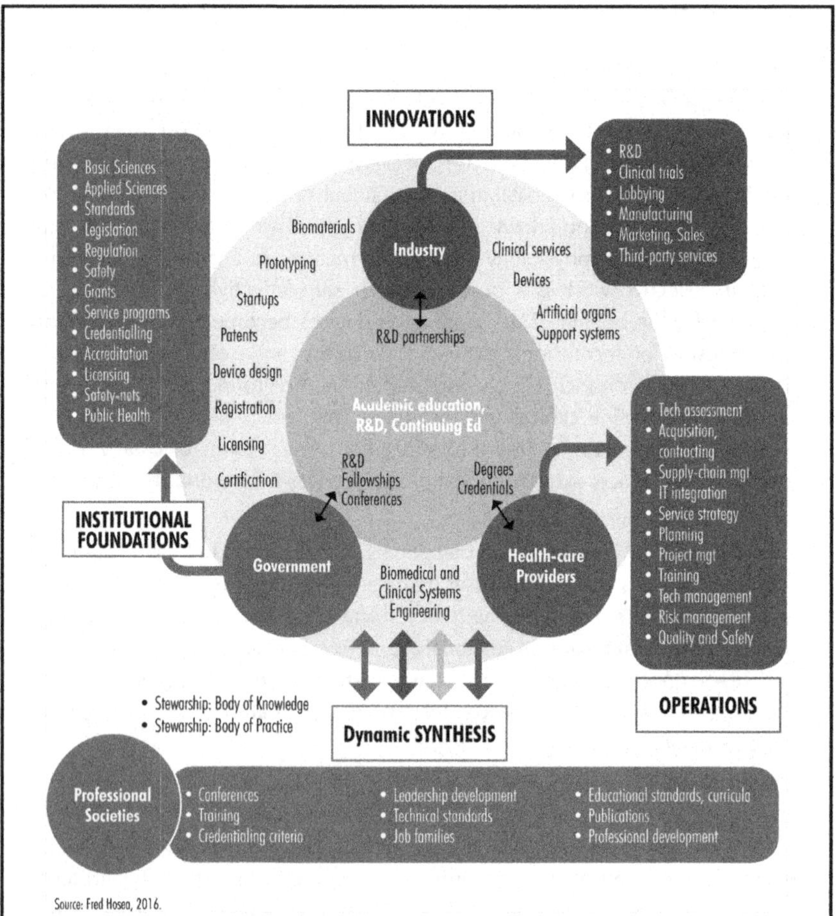

Source: Fred Hosea, 2016.

Figure 3 Roles of biomedical engineers

Biomedical engineering tasks and responsibilities defined

During the 2015 Global Clinical Engineering Summit *(2015)(30)* the following roles and responsibilities were defined, as well as the subspecialisms of BME (shown in Table 1).

Applying knowledge of engineering and technology to health-care systems to optimize and promote safer, higher quality, effective, affordable, accessible, appropriate, available, and socially acceptable technology to populations served by:

1. Advancing health and wellness using technologies for prevention, diagnosis, treatment, rehabilitation and palliative care across all levels of the health-care delivery;

Table 1 Subspecialisms of BME

Research and development	Rehabilitation	Application and operation: clinical engineering
Biomechanics	Artificial organs	Technology management
Biomaterials	Neural engineering	Quality and regulatory assurance
Bioinformatics	Tissue engineering /regenerative	Education and training
Systems biology	Mechatronics	Ethics committee, clinical trials
Synthetic biology	Assistive devices and software	Disaster preparedness
Bionics	Prosthetics	e-health (telemedicine, m-health)
Biological engineering		Wearable sensors/products
Nanotechnology		Health economics
Genomics		Health systems engineering
Population health/data analytics		Health technology assessment/ evaluation
Epidemiology (computational)		Health informatics
Intellectual property/innovation		Service delivery management
Theranostics		Field service support
Biosignals		Security/privacy/cybersecurity
		Forensic engineering/ investigation
		Manufacturing QMS, GMP
		Medical imaging
		Project management
		Robotics
		Virtual environments
		Risk management
		EMI/EMC compliance
		Technology Innovation strategies
		Population- and community-based needs assessment
		Engineering asset management
		Environmental health
		Systems science

2. Innovating, designing, developing, regulating, managing, assessing, installing, and maintaining such technologies for their safe and cost effective use throughout their life cycle;
3. Applying engineering principles and design concepts to medicine and biology for the pursuit of new knowledge and understanding at all biological scales;
4. Designing devices, software, processes and techniques to be used in wellness and health care, including consumables, artificial organs and prosthesis, diagnostic and therapeutic instrumentation and related systems such as magnetic resonance imaging, and devices for automating insulin injections or controlling body functions;
5. Designing, developing and managing technologies used to promote and support life quality and longevity, including assistive technologies and technologies for monitoring or rehabilitating activities of daily living; such as wheel chairs, prosthesis leg, hearing aid and personal emergency response systems;
6. Designing, developing and managing technologies for focus areas such as reproductive, maternal, neonatal, and child health;
7. Designing, developing and managing systems for optimal sustained health-care operations in both resource-scarce and well resourced settings as well as during challenging events such as disasters; and
8. Designing, developing and applying safety programme methodologies to mitigate risks when dealing with medical devices and procedures throughout their life cycle. Including biosafety and environmental health such as waste disposal and personal radiation protection.

Health-care technologies include: health, wellness and rehabilitation products and systems, artificial biological structures, organs, and prostheses, instrumentation, software and multi-technology systems.

Biomedical engineering

Trained and qualified BME professionals are required within health-care systems in order to design, evaluate, regulate, acquire, maintain, manage and train on safe medical technologies. The BME profession, however, is often not included in the official definitions of the health workforce or policy frameworks, and this absence significantly impairs the advancement and sustainability of these health-care systems, even in settings with available resources.

The International Labour Organization (ILO) manages the International Standard Classification of Occupations (ISCO), which organizes the tasks and duties of jobs, with the objective of having international reporting and statistical data of occupations and serves to enhance national and regional classification of occupations. The current system, ISCO-08, classifies BME professionals as a part of Unit Group 2149 "Engineering Professionals not Elsewhere

Classified."*(31)* The classification of and statistics regarding biomedical engineering professionals by the ILO are undergoing formal review as part of the ILO's 10-year cycle for classifying the world's professions.

According to the current ISCO-O8, from 2008, "biomedical engineering" professionals are considered to be an integral part of the health workforce alongside those occupations classified in Sub-major Group 22: Health Professionals. It is important to recognize that although ISCO-O8 has "noted" biomedical engineers as an integral part of the health workforce, the profession has not yet been independently re-classified as a specialized type of engineering. Specialized classification has been requested in the past, but the small number of countries with biomedical engineering professionals and the shortage of professionals at that time made this impossible.

In the past, biomedical engineering professionals have worked "in the shadow" of more recognized professions, such as doctors and nurses. This has been in part due to lack of official research and dissemination of information about the presence and essential value of biomedical engineering professionals worldwide. This gap in acknowledgement and inclusion of BME within the health-care workforce, and the lack of current data on the profession, both urgently need to be addressed, to ensure that the health-care sector has the necessary mix of professionals to guide the dynamic changes in science, technology and services in the 21st century.

In recent years, numbers have increased significantly, with documentation showing biomedical engineering professionals in 126 out of 194 WHO Member States (64%), and the scope and depth of BME expertise increasing with a growing presence of the profession globally in health-care systems. In 2012, data from the United States Bureau of Labor Statistics and other governmental agencies ranked BME as the second best profession based on five criteria: physical demand, work environment, income, stress and hiring prospects.*(32)*

Should ISCO re-classify biomedical engineering professionals along with other healthcare professions in 2018 or future years, more statistics and data will be available at the country level, and formal recognition processes could be initiated by member countries to recognize BME as a profession within their national or regional labour organizations and ministries.

According to the International Federation of Medical and Biological Engineers (IFMBE) — an NGO in official relations with WHO that represents professional and scientific interests of 59 national member societies from around the world - BME is defined as follows:

> *Medical and biological engineering integrates physical, mathematical and life sciences with engineering principles for the study of biology, medicine and health systems and for the application of technology to improving*

health and quality of life. It creates knowledge from the molecular to organ systems levels, develops materials, devices, systems, information approaches, technology management, and methods for assessment and evaluation of technology, for the prevention, diagnosis, and treatment of disease, for health care delivery and for patient care and rehabilitation.(33)

In order to update these issues on the classification, statistics and recognition of the BME profession, WHO began efforts to track the presence of BME professionals and technicians worldwide. In 2009, WHO launched "Biomedical engineering global resources,"*(34)* a WHO programme to gather information on academic programmes, professional societies and the status of BME worldwide; the results of which are presented in this publication.

Global Strategy on Human Resources for Health: Workforce 2030

In May 2014, the Sixty-seventh World Health Assembly adopted resolution WHA67.24 on Follow-up of the Recife Political Declaration on Human Resources for Health: renewed commitments towards universal health coverage. In paragraph 4(2) of that resolution, Member States requested the Director-General of WHO to develop and submit a new global strategy for human resources for health (HRH) for consideration by the Sixty- ninth World Health Assembly. A summary of the Global Strategy on Human Resources for Health: Workforce 2030 can be found in Annex 5.*(35)*

The goal of the global strategy is: to improve health, social and economic development outcomes by ensuring universal availability, accessibility, acceptability, coverage and quality of the health workforce through adequate investments to strengthen health systems, and the implementation of effective policies at national, a regional and global levels.

The global strategy, includes all cadres involved in delivery of health services, as described section 16 of the strategy:

*16. "This is a cross-cutting agenda that represents the critical pathway to attain coverage targets across all service delivery priorities. It affects not only the better known cadres of midwives, nurses and physicians, but **all health workers**, from community to specialist levels, **including but not limited to:** community-based and mid-level practitioners, dentists and oral health professionals, hearing care and eye care workers, laboratory technicians, **biomedical engineers**, pharmacists, physical therapists and chiropractors, public health professionals and health managers, supply chain managers, and other allied health professions and support workers. The Strategy recognizes that diversity in the health workforce is an opportunity to be harnessed through strengthened collaborative approaches to social accountability, inter-professional education and practice, and closer integration of the health and social services workforces to improve long-term care for ageing populations."*

The global strategy requests the support of professional organizations to regulate the workforce competency as described below:

36. "Professional councils to collaborate with governments to implement effective regulations for improved workforce competency, quality and efficiency. Regulators should assume the following key roles: keep a live register of the health workforce; oversee accreditation of pre-service education programmes; implement mechanisms to assure continuing competence, including accreditation of post-licensure education providers; operate fair and transparent processes that support practitioner mobility and simultaneously protect the public; and facilitate a range of conduct and competence approaches that are proportionate to risk, and are efficient and effective to operate.(49) Governments, professional councils and associations should work together to develop appropriate task-sharing models and inter-professional collaboration, and ensure that all cadres with a clinical role, beyond dentists, midwives, nurses, pharmacists and physicians, also benefit in a systematic manner from accreditation and regulation processes".

The United Nations High-Level Commission on Health Employment and Economic Growth was established in March 2016, to recommend the creation of 40 million jobs in the health and social sector, particularly in low- and middle-income countries (LMIC), for 2030. It made 10 recommendations to transform the health workforce for the SDG era. These recommendations include job creation, gender and women's rights, education and training skills, health service delivery and organization, partnership and collaboration, and data, information and accountability (further information can be found in Annex 6). It is important to note the Commission already acknowledges the role of technological advances related to medical devices in encouraging economic development and supporting the health sector:

> *"The innovation and diversification pathway illustrates how some countries have invested in their health sector specifically to promote economic growth. The health sector has been driving technological innovations in many areas, including genetics, biochemistry, engineering and information technology. Exports of pharmaceuticals, equipment and medical services have also been an important driver of growth in many countries."*
>
> ***http://www.who.int/hrh/com-heeg/reports/en/***

The commission's recommendations are aligned with the SDGs. The specific target and goals related to this publication on human resources for medical devices are:

SDG 3: Good health and well-being

Target 3.c: Substantially increase health financing and the recruitment, development, training and retention of the health workforce in developing countries, especially in least developed countries and small island developing States.

SDG 4: Quality education

Target 4.3: By 2030, ensure equal access for all women and men to affordable and quality technical, vocation and tertiary education, including university.

Target 4.b: By 2020, substantially expand globally the number of scholarships available to developing countries, in particular least developed countries, small island developing States and African countries, for enrolment in higher education, including vocational training and information and communications technology, technical, engineering and scientific programmes, in developed countries.

Questions to Consider

Observations and important points:

The World Health Organization presents reports that are applicable across the globe. Thus, some terminology and general practices may vary from the profession of clinical engineering in the United States. While the entire document is not included in this text, the publication is readily available on the Internet.

- Begin with the definitions section. Note that the publication utilizes the term *biomedical engineer* yet acknowledges the *clinical engineer* as a related profession. Review these definitions, and explore the alignment with the descriptions commonly found in the U.S. health care system. Are these differences a challenge?
- Explore Figure 1. This illustration is unique as it documents the touchpoints associated with medical devices. How does this approach expand the understanding of the role of a clinical engineer?
- Look at the Responsibilities section. Are these characterizations well-aligned with the clinical engineer role this book has described? Does the use of the term *biomedical engineer* in this context hinder understanding or application of the description?
- Search the Internet to find the International Federation of Medical and Biological Engineering—Clinical Engineering Division (IFMBE-CED). How can this group advance understanding of the profession?
- How can the World Health Organization improve awareness for the role of engineers (biomedical, clinical, or otherwise named) in the delivery of safe and effective health care?
- Does access to technical training and higher education influence the delivery of safe and effective health care? Does this publication characterize the tight relationship between education and population wellness?

Article 3: Nader, R., Ralph Nader's Most Shocking Expose. Ladies Home Journal. April 24, 1970. 176–179

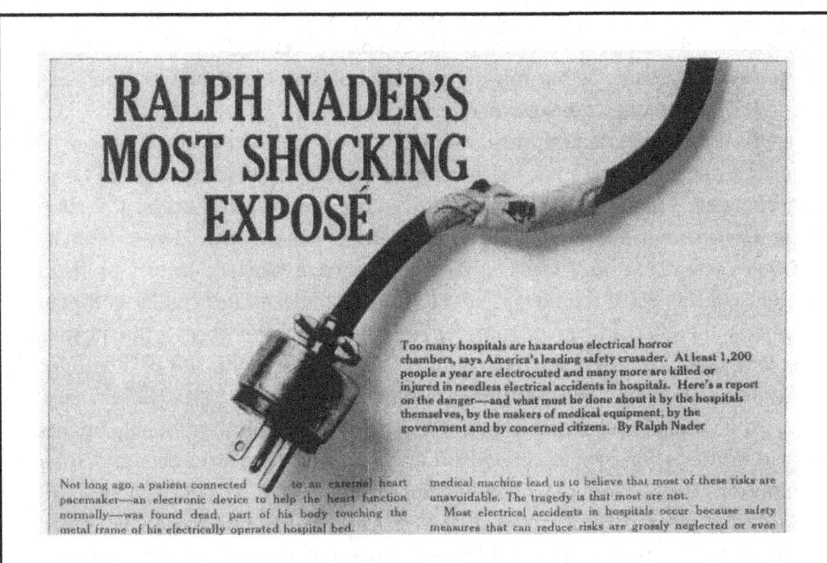

RALPH NADER'S MOST SHOCKING EXPOSÉ

Too many hospitals are hazardous electrical horror chambers, says America's leading safety crusader. At least 1,200 people a year are electrocuted and many more are killed or injured in needless electrical accidents in hospitals. Here's a report on the danger—and what must be done about it by the hospitals themselves, by the makers of medical equipment, by the government and by concerned citizens. By Ralph Nader

Not long ago, a patient connected to an external heart pacemaker—an electronic device to help the heart function normally—was found dead, part of his body touching the metal frame of his electrically operated hospital bed.

medical machine lead us to believe that most of these risks are unavoidable. The tragedy is that most are not.

Most electrical accidents in hospitals occur because safety measures that can reduce risks are grossly neglected or even

Too many hospitals are hazardous electrical horror chambers, says America's leading safety crusader. At least 1,200 people a year are electrocuted and many more are killed or injured in needless electrical accidents in hospitals. Here's a report on the danger—and what must be done about it by the hospitals themselves, by the makers of medical equipment; by the government and by concerned citizens. By Ralph Nader

Not long ago, a patient connected *M to an external* heart pacemaker—an electronic device to help *the heart* function normally—was found dead, part of his body touching the metal frame of his electrically operated hospital bed.

In another hospital, a resident physician was discovered slumped lifeless beside a stainless steel table. He had been electrocuted when he touched an ungrounded oscilloscope (an instrument that monitors the heart pacemaker) and the table at the same time.

In yet another hospital, a patient suddenly became rigid during a routine diagnostic procedure, warning personnel to cut the electric power of an instrument that was sending potentially lethal currents into his heart. Fortunately, the patient survived.

In a fourth hospital, an electrical switch broke and a patient was crushed to death by a descending X-ray machine.

And in a speech last November 16, Roger O. Egeberg. M.D., Assistant Secretary for Health and Scientific Affairs of the U.S. Department of Health, Education and Welfare, described another hospital tragedy: "Not long ago," Dr. Egeberg noted, "a woman in her midsixties entered a hospital in metropolitan Washington, D.C., for routine thyroid gland surgery. When the operation was completed and the patient was being sutured, the physician turned off the anesthesia machine. An explosion occurred, possibly caused by an electrical spark. Within four and a half hours the patient was dead as a result of the injuries she sustained in the blast."

Paradoxically, medical instruments that have brought hope of longer life to thousands of people have also increased a thousandfold the risks to hospital patients. "Life-saving" electrical devices used in hospitals across the country electrocute an average of three patients a day, at the lowest estimate. Other patients die as a result of electrical burns, explosions or loss of instrument control. Since the advent of the heart pacemaker and cardiac catheterization—the insertion of a catheter, or tube, into the heart—the hospital environment has become so dangerous that today it is the site of more electrical accidents than any industry except mining.

Spectacular advances in medical technology have unquestionably opened new horizons for people suffering from heart and lung disorders and other diseases. Those who may benefit—for example, the 20,000 patients who receive implanted heart pacemakers each year—may understandably be willing to hazard risks in hope of staying alive. But the myths of the medical machine lead us to believe that most of these risks are unavoidable. The tragedy is that most are not.

Most electrical accidents in hospitals occur because safety measures that can reduce risks are grossly neglected or even unknown among hospital staffs; because complex and highly dangerous equipment is installed in hospitals that have primitive wiring systems, and the equipment is operated by untrained personnel; and because machines that reach inside a human being and touch his heart are less well tested than plumbing devices in our bathrooms.

These accidents often occur because manufacturers design dangerous devices without making them fail-safe against even the most common mistakes of operators. The real risk for a hospital patient may be considerably less than esoteric. It may be the risk that hospital staff will decide to use a frayed electrical cord one more time, or, for the heart patient with an external pacemaker, that he will be placed in an electrically operated bed—a highly dangerous but common occurrence. Or instead of employing a qualified biomedical engineer, a hospital administrator may ask the building electrician to install complex new equipment.

Most of these instances of negligence remain hidden by the fact that physicians and hospitals habitually report deaths by electrocution as "cardiac arrest."

Electrical gadgetry and the accompanying hazards of electric shock are everywhere in our environment—in our homes, schools and offices. When a young guitarist is electrocuted by his instrument, or when a priest is killed by an electrically operated weight reducer—two incidents recently reported in the press—we want to know what went wrong. Was the guitar defective? Was the wiring bad? Did the victim use the machine improperly? Unfortunately, these questions are not often asked in hospitals, where at the very least 1,200 Americans are electrocuted annually during routine diagnostic and therapeutic procedures.

We do not even have a clear idea of the number of hospital fatalities caused by electric shock. Medical engineers such as Professor Hans von der Mosel, co-chairman of the Subcommittee on Electrical Safety of the Association for the Advancement of Medical Instrumentation and safety consultant to New York City's Health Services Administration, believe that the number might be I 0 times as high as the conservative estimate of 1,200. Yet most of these deaths could have been prevented by adequate safety measures.

For the patient whose heart is made accessible to electric current through electrodes and catheters, merely touching the frame of a hospital bed, especially an electrically operated bed, may prove fatal. This happened to a 52-year-old man who was connected to an external pacemaker by means of a catheter inserted into the heart itself. Someone had attached to the pacemaker an ungrounded extension cord that eliminated the instrument's grounding system. When a current leaked from the pacemaker, as it frequently does, it passed through the catheter electrode into the patient's heart, then through the part of his body in contact with the grounded electrically elevated bed.

The death could have been prevented in at least three ways: if hospital staff had not attached an ungrounded extension cord to the pacemaker (extension cords should never be used with such equipment); if the patient had not been placed in an electrically operated bed; if the pacemaker had carried a device that limited the current in the patient's circuit to a safe level.

This death was investigated because it was the third such fatality in less than two months at that hospital. It is possible, even likely, that the other deaths, which were not investigated, were also due to electrocution. But most such deaths are not reported. Almost invariably, when electrocutions happen during diagnostic procedures in which the patient is hooked up to electronic systems, the deaths are listed as cardiac arrests. Without engineering analysis, it is difficult, to tell whether a patient died of his disease or of a shock caused by the equipment. To protect themselves against malpractice suits, physicians and hospitals avoid such investigations, and many hazards go undetected and uncorrected. There have been few lawsuits over these deaths, and thus the hazards have been little publicized. Insurance companies that make studies of electrical hazards have not alerted the public to the dangers or to the incidence of death. Statistics have hidden the fact that a shock-hazard epidemic of critical proportions exists in our hospitals.

The hazards of electrical devices are not limited to delicate equipment such as the heart pacemaker. Routine electrical equipment may also cause death. Take the case of the patient who was squeezed to death when the switch controlling the X-ray machine's vertical movement failed while the machine was being lowered over him. He died before the technician could open the circuit breaker located some distance away.

What caused the switch to fail? A broken contact blade that shorted a circuit. The break had probably existed for some time. If the hospital had conducted regular equipment checks, the fault would probably have been discovered and corrected. In addition, precious time was lost because the main switch, which cuts all power to the instrument, was not easily accessible. Finally, the circuit breakers were not clearly marked, and the technician opened three different electrical circuits before he found the right one.

Nor are electrical accidents limited to patients. A young Canadian physician nearly died of electric shock when he pressed the discharge button on a defibrillator. This machine, used for correcting uncoordinated heartbeat, is inherently dangerous because it is designed to deliver a high-energy shock. Examination revealed that the ground wire in the three-prong plug had been broken, presumably when someone attempted to force the plug into a two-hole socket. Thus current was released—first into the chassis of the machine, then into the physician.

Inexcusable negligence

Some fatalities are caused by inexcusable negligence. Many devices are used with adapter plugs that don't ensure grounding. That is what happened with the hospital doctor who was found dead, the metallic switch of the oscilloscope in his right hand, his left hand touching a metal drawer of the stainless steel table on which the instrument was standing. A device in the power supply circuit of the oscilloscope had shorted, shooting 300 volts into the cabinet of the instrument. The oscilloscope should have been grounded through the grounding prong of the three-prong connector; instead a three- to-two-prong adapter (called a "cheater adapter") was in use. In this case, the adapter was completely unnecessary to connect the instrument, but the instrument was not designed to prevent the mistake. Because it was ungrounded, and because the doctor was touching a grounded steel table, the current passed through his right arm, through his trunk, heart and left arm into the grounded table.

Physicians and hospital personnel have been aware for some time of the hazards of electrically ignited explosions and external electric shock.

Some progress has been made in reducing the danger of explosions in operating rooms by employing standard safety precautions and, in a few hospitals, by eliminating flammable anesthetics. But there is little if any protection against a newer hazard—internal shock. Catheters, electrodes and probes have opened pathways to the heart through which very small accidental currents can kill a patient. A shock of 20 microamps across the heart can cause fibrillation, which after one minute results in irreversible brain damage and after three minutes, in death. At the surface of the body, a shock must be a thousand times greater to produce fibrillation.

Six Ways to Make Your Hospital Safer Electrically

1. Organize a citizens' group to investigate the administrator of your community hospital's electrical safety system. If possible, take an electrical engineer with you, or have one brief you on important questions to ask. Does the hospital have proper wiring? Does the hospital utilize the services of biomedical engineers? Are devices and device systems tested before they are hooked up to a patient? What are the provisions for testing new equipment?
2. Citizens and community groups can demand investigation of hospital fatalities. Find out who monitors accidents in your community hospital. When accidents are attributed to "cardiac arrest," was that the real cause of death? Are electrical systems always tested after deaths and injuries occur that could be attributed to electrical or equipment failure?
3. Is there an electrical device safety committee at your hospital? Organize a group of concerned citizens to meet with hospital review committees to ask what precautions are taken in the use of electrical devices.
4. Ask the company that insures your community hospital for statistics on electrical accidents at the hospital.
5. Urge your newspaper to make a thorough investigation of electrical devices in local hospitals and to publicize any particular problems.
6. If you are a professional engineer, take the lead in exposing and correcting electrical problems in local hospitals.

If properly grounded, most devices are safe when used by themselves. But most of the time, the patient is connected not to one but to several electrical devices. In addition, he may touch any number of other electric appliances—bed, radio, television, clock, lamp. He may also come in contact with routine equipment, such as portable X-ray machines, physiotherapy apparatus and respirators. In such an environment, the risk is extremely great that a stray electrical current will complete a circuit to ground through the patient. Most electrocutions happen in just this way. Prevention of death or injury from internal shock requires expert planning, sophisticated wiring systems, and careful, constant testing.

Few hospitals, even the newer ones, have adequate electric wiring systems. Most need extensive modernization to provide a safe environment for new electrical devices that are in widespread use. Electrical overloading is common. Many hospital outlets are incorrectly wired or provide no ground contact. In most cases these outlets were installed by hospital electricians when equipment appeared with three wires. As long as the plugs went in, the electrician believed his job was done.

Only three hospitals in the country have biomedical engineers on their staffs to supervise the operation and maintenance of complex machines: Downstate Medical Center in New York City; Sinai Hospital in Baltimore; and Charles S. Wilson Hospital in Johnson City, N.Y. Most hospitals simply turn over the apparatus to a staff physician who may have worked with electronic equipment. Hospitals do not yet have electrical device safety committees comparable to drug safety committees, although the two hazards are equally great. Few physicians who deal with these devices know as much about the concepts behind them or about their use as they know about pharmacology. Yet for years physicians operated these devices without recognizing either their potential hazards or the actual fatalities they caused. Countless deaths attributed to cardiac arrest are now believed to have been caused by internal electric shock. Even now that there is greater understanding of the risks posed by the new hospital environment, precautionary measures are inadequate.

While inadequate hospital facilities and errors in using the machines are leading causes of accidents, mechanical defects also play a part in imperiling patients' lives. One medical engineer, Seymour Ben-Zvi, tested several thousand instruments at Down- state Medical Center in New York City. He reported that 40 percent were defective. Every one of the 10 defibrillators he tested contained defects. One was capable of discharging high voltage into a patient before the physician signaled for it. Such a defect could kill both patient and physician. Another instrument had what the manufacturer thought was an insulator; it was actually a good conductor of electricity—a potentially fatal flaw that should have been discovered through testing. (The testing program at Downstate began in 1956, and Ben-Zvi states that most manufacturers now agree to correct defects found.)

C. W. Walter, a clinical professor of surgery at Harvard Medical School, has reported that two prominent firms are now selling highly dangerous machines. Poor circuit design is a common criticism, and many devices have problems stemming from high leakage of current, problems often revealed only through the death of a patient. Some manufacturers offer to replace equipment; they cannot replace a dead person.

Toilets and pacemakers

Mrs. Virginia Knauer. President Nixon's Assistant for Consumer Affairs, has pointed out that toilet valves must pass several preclearance tests before they are installed in our bathrooms, but a pacemaker that is inserted into our hearts need not be tested at all. Heart pacemakers, artificial kidneys, hip pins and respirators—none are subject to standard inspection or regulation—as are drugs, for example.

Manufacturing of medical devices is a $500-million-a- year industry engaged in by more than 1,000 firms. Without regulations or standards, there has been little impetus for these firms to standardize their products. Manufacturers' resistance to standardization has created an unnecessary hazard, since each hospital must sort out discrepancies in connectors and devise a system to prevent hazardous currents from being applied to helpless patients. Generally, the manufacturer considers his product a separate unit rather than part of a total treatment system, although a device is rarely used by itself.

In designing instruments, manufacturers almost totally ignore the ease with which mistakes can be made in the hospital environment, where personnel are often hurried, strained or tired, and untrained in the use of the equipment. Fatal errors are made that could be prevented by safer design.

Often the grounding devices furnished with electrical equipment are weak, easily broken and not designed for rough handling. They are not remotely foolproof, not fail-safe and not even reliable. Cords and plugs, the most vulnerable part of the electrical safety system, are usually "totally inadequate," according to Professor Walter. On occasion, a complex and expensive piece of equipment is equipped with a cheap, inefficient plug.

Fatally for patients and staff, manufacturers often assume that users have technical competence, which they almost universally lack. Instruction booklets, labels, foolproofing and protection devices are far inferior to what is needed.

One respected independent testing agency that has begun to test and evaluate medical equipment reports an "appalling" number of defective instruments. Research at the Emergency Care Research Institute of Philadelphia revealed, for example, defective respirators that were "totally unable to support respiration." The Food and Drug Administration has recalled a number of these devices.

Dr. Joel J. Nobel, ECRI's director of research, says that "the number of life-threatening defects is truly appalling. Most are basic design deficiencies."

ECRI is a nonprofit organization supported by government agencies, hospitals and private contributions. No staff member receives consulting fees from the health devices industry. ECRI findings indicate that, in the absence of objective testing and evaluation, unsafe equipment is being used in hospitals that are unequipped to pretest it.

Hospital associations in three regions—California, Texas and New England—are in the process of setting up medical product information exchange systems. Central testing programs to serve all hospitals in a region are much more feasible than tests conducted in individual hospitals, but such programs have yet to get underway.

At present, there are no government regulations requiring premarket clearance or standards to ensure the safety and performance of certain medical devices, such as catheters, pacemakers, diathermy machines and bone pins. During the past several years, efforts to bring new devices under regulations have failed. Presidents Kennedy, Johnson and Nixon have supported regulations and minimum standards for medical devices. In September 1970 a study group appointed by President Nixon and headed by Dr. Theodore Cooper, Director of the National Heart and Lung Institute of the National Institutes of Health, recommended legislation to regulate these devices.

A bill has been introduced by Congressman Thomas Foley (D., Wash.) to establish regulations and standards for devices not covered by present law. This bill was originally proposed in 1969, but no action has yet been taken. Legislation has been stymied in part by claims that standards for such instruments are difficult to set. But the failure of physicians to publicize the real extent of the hazards is the reason why the need for legislation has been unnoticed.

Pretesting of these devices by independent testing agencies and establishment of uniform government standards will help ensure that the instruments are safe, that they are fail-safe and that they assume much less knowledge and expertise on the part of the typical hospital employee who runs them.

But beyond government standards, what is needed is greater vigilance by hospitals and physicians. In the absence of trained personnel, adequate electrical systems and rigid inspection and testing, even the best designed machine may become a killer. Unfortunately, there is little indication, on a broad scale, that hospitals and physicians are prepared to make a major commitment to electrical safety. Instead, there is every indication that accidents are occurring more frequently. The public may well ask where the electric safety committees in hospitals are, or the services of biomedical engineers.

Where are the research grants to study questions of safety? Where is the leadership of medical organizations that should be demanding safety from manufacturers and help in ensuring safety from governments? I do not believe the public should have to accept the response one physician made to the problem of hospital safety: that after all. most electrical accidents occur in the home.

It is true that there is too little understanding of electrical hazards. The use of two-prong plugs (without a third grounding wire) is a simple hazard that continues to exist in many homes and other buildings. The naivete of physicians who use intricate devices is undoubtedly shared by many other people who do not understand when or why electrical devices can be hazardous. The housewife who simultaneously touches a toaster and a refrigerator handle and receives a shock usually lives to return the toaster, or change the wiring, or complain to the manufacturer. The heart patient who receives the same kind of shock is not so fortunate.

If we have the technology to stimulate the heart, to sustain life and to probe the innermost regions of the body, we also have the means to make devices that are safe from human error. The unprecedented hope offered by new medical technology does not need to be accompanied by unprecedented risk. Such avoidable tragedies in our hospitals will not be stopped until manufacturers recognize the limitations of the personnel who use their devices, and until users demand that safety be built into the devices. Dangers that have been veiled as unavoidable risks, or risks inherent in the condition of the patient, must be exposed. Until they are, new medical devices will continue their Jekyll-and-Hyde role—they are life-giving devices for some, but death machines for others.

Source: *Reprinted courtesy of Meredith Corporation, publisher of* Ladies' Home Journal ® *magazine.*

Questions to Consider

- Begin by exploring the scholarly value of this periodical. Did the readership of *Ladies' Home Journal* influence the understanding and reaction to the "report"? Who is the author? What is the author's expertise in this area?
- Can the reader identify how the author determined the number of deaths per day/per year as reported in the article? Might this value be accurate? Is the accuracy important?

- Some experts were cited in the article, including a representative from ECRI. Examine the history of ECRI and how they may have information about the danger of medical devices.
- The author states that medical device manufacturers "design dangerous devices" with "basic design deficiencies". Why might the author have described technology in this way? Was the equipment significantly primitive? Explore devices available in the early 1970s, and assess this premise.
- The author's recommendations include the importance of the "services of biomedical engineers," yet one must assess this recommendation that was made at almost the same time as the National Academy of Engineering report that described only three biomedical engineering academic programs established at the time of publication (Johns Hopkins, Ohio State University, and University of Wisconsin). Could the number of academic graduates been satisfactory for the number of hospitals?
- This article is often described as the seminal event that launched hospital safety awareness and the push for technicians in hospitals, though the number and role of baccalaureate or graduate degree engineers would expand at a much slower pace. Many technicians arose from the maintenance department, initially trained as electricians or plumbers. How has this affected the support of medical devices in today's hospitals?
- Explore the 1976 federal legislation that added Medical Device Amendments to the Food and Drug Administration Act that established classifications of medical devices and regulated manufacturing practices. The Food and Drug Administration has an excellent resource on the Internet associated with the history of medical device regulation and oversight in the United States. Is there a probable link between Ralph Nader's expose and the regulation?

Article 4: Building a better delivery system: A new engineering/ health care partnership. National Academy of Engineering and Institute of Medicine (2005)

Executive Summary

American medicine defines the cutting edge in most fields of clinical research, training, and practice worldwide, and U.S.-based manufacturers of drugs, medical devices, and medical equipment are among the most innovative and competitive in the world. In large part, the United States has achieved primacy in these areas by focusing public and private resources on research in the life and physical sciences and on the engineering of devices, instruments, and equipment to serve individual patients.

At the same time, relatively little technical talent or material resources have been devoted to improving or optimizing the operations or measuring the quality and productivity of the overall U.S. health care system. The costs of this collective inattention and the failure to take advantage of the tools, knowledge, and infrastructure that have yielded quality and productivity revolutions in many other sectors of the American economy have been enormous. The $1.6 trillion health care sector is now mired in deep crises related to safety, quality, cost, and access that pose serious threats to the health and welfare of many Americans (IOM, 2000, 2001, 2004a,b,c).

One need only note that: (1) more than 98,000 Americans die and more than one million patients suffer injuries each year as a result of broken health care processes and system failures (IOM, 2000; Starfield, 2000); (2) little more than half of U.S. patients receive known "best practice" treatments for their illnesses and less than half of physician practices use recommended processes for care (Casalino et al., 2003; McGlynn et al., 2003); and (3) an estimated thirty to forty cents of every dollar spent on health care, or more than a half-trillion dollars per year, is spent on costs associated with "overuse, underuse, misuse, duplication, system failures, unnecessary repetition, poor communication, and inefficiency" (Lawrence, in this volume). Health care costs have been rising at double-digit rates since the late 1990s— roughly three times the rate of inflation—claiming a growing share of every American's income, inflicting economic hardships on many, and decreasing access to care. At the same time, the number of uninsured has risen to more than 43 million, more than one-sixth of the U.S. population under the age of 65 (IOM, 2004a).

With support from the National Science Foundation (NSF), National Institutes of Health (NIH), and Robert Wood Johnson Foundation, the National Academy of Engineering (NAE) and Institute of Medicine (IOM) of the National Academies convened a committee of 14 engineers and health care professionals to identify engineering tools and technologies that could help the health system overcome these crises and deliver care that is safe, effective, timely, patient-centered, efficient, and equitable—the six quality aims envisioned in the landmark IOM report, *Crossing the Quality Chasm* (Box ES-1).

The committee began with the expectation that systems- engineering tools that have transformed the quality and productivity performance of other large-scale complex systems (e.g., telecommunications, transportation, and manufacturing systems) could also be used to improve health care delivery. The particular charge to the committee was to identify: (1) engineering applications with the potential to improve health care delivery in the short, medium, and long terms; (2) factors that would facilitate or inhibit the deployment of these applications; and (3) priorities for research and education in engineering, the health professions, and related areas that would contribute to rapid improvements in the performance of the health care delivery system. The committee held three intensive workshops with experts from the engineering, health, management, and social science communities. The presentations by these experts can be found in Part 2 of this volume.

BOX ES-1 Six Quality Aims for the 21st Century Health System

Six aims for improvement to address key dimensions in which today's health care system functions at far lower levels than it can and should. Health care should be:

- Safe—avoiding injuries to patients from the care that is intended to help them.
- Effective—providing services based on scientific knowledge to all who could benefit and refraining from providing services to those not likely to benefit (avoiding underuse and overuse, respectively).
- Patient-centered—providing care that is respectful of and responsive to individual patient preferences, needs, and values and ensuring that patient values guide all clinical decisions.
- Timely—reducing waits and sometimes harmful delays for both those who receive and those who give care.
- Efficient—avoiding waste, including waste of equipment, supplies, ideas, and energy.
- Equitable—providing care that does not vary in quality because of personal characteristics such as gender, ethnicity, geographic location, and socioeconomic status.

Source: IOM, 2001, pp. 5–6.

Engineering-Health Care Partnership

This report provides a framework and action plan for a systems approach to health care delivery based on a partnership between engineers and health care professionals. The goal of this partnership is to transform the U.S. health care sector from an underperforming conglomerate of independent entities (individual practitioners, small group practices, clinics, hospitals, pharmacies, community health centers, et al.) into a high-performance "system" in which every participating unit recognizes its dependence and influence on every other unit. The report describes the opportunities and challenges to harnessing the power of systems-engineering tools, information technologies, and complementary knowledge in social sciences, cognitive sciences, and business/ management to advance the six IOM quality aims for a twenty-first century health care system.

This NAE/IOM study attempts to bridge the knowledge/ awareness divide separating health care professionals from their potential partners in systems engineering and related disciplines. After examining the interconnected crises facing the health care system and their proximate causes (Chapter 1), the report presents an overview of the core elements of a systems approach and puts forward a four-level model— patients, care teams, provider organizations, and the broader political-economic environment—of the structure and dynamics of the health care system that suggests the division of labor and interdependencies and identifies levers for change (Chapter 2).

In Chapters 3 and 4, systems-engineering tools and information/communications technologies and their applications to health care delivery are discussed. These complementary tools and technologies have the potential of improving radically the quality and productivity of American health care. Structural, economic, organizational, cultural, and educational barriers to using systems tools and information/ communications technologies, and recommendations for overcoming these barriers follow. In Chapter 5, the committee proposes a strategy for building a vigorous partnership between engineering and health care through cross- disciplinary research, education, and outreach.

Systems-Engineering Tools for Health Care Delivery

Systems-engineering tools have been used in a wide variety of applications to achieve major improvements in the quality, efficiency, safety, and/or customer-centeredness of processes, products, and services in a wide range of manufacturing and services industries. The health care sector as a whole has been very slow to embrace them, however, even though they have been shown to yield valuable returns to the small but growing number of health care organizations and clinicians that have applied them (Feistritzer and Keck, 2000; Fone et al., 2003; Leatherman et al., 2003; Murray and Berwick, 2003). Statistical process controls, queuing theory, quality function deployment, failure-mode effects analysis, modeling and simulation, and human-factors engineering have been adapted to applications in health care delivery and used tactically by clinicians, care teams, and administrators in large health care organizations to improve the performance of discrete care processes, units, and departments.

However, the strategic use of these and more information- technology-intensive tools from the fields of enterprise and supply-chain management, financial engineering and risk analysis, and knowledge discovery in databases has been limited. With some adaptations, these tools could be used to measure, characterize, and optimize performance at higher levels of the health care system (e.g., individual health care organizations, regional care systems, the public health system, the health research enterprise, etc.). The most promising systems-engineering tools and areas of associated research identified by the committee are listed in Table ES-1.

Although data and associated information technology needs do not present significant technical or cost barriers to the tactical application of systems-engineering tools, there are significant structural, technical, and cost-related barriers at the organizational, multi-organizational, and environmental levels to the strategic implementation of systems tools. The current organization,

Table ES-1 Systems Engineering Tools and Research for Health Care Delivery

Tools/Research Areas	Levels of Application			
	Patient	Team	Organization	Environment
SYSTEMS-DESIGN TOOLS				
Concurrent engineering and quality function deployment		X	X	
Human-factors tools	X	X	X	X
Failure mode effects analysis		X	X	
SYSTEMS-ANALYSIS TOOLS				
Modeling and Simulation				
Queuing methods	X	X	X	
Discrete-event simulation	X	X	X	X
Enterprise-Management Tools				
Supply-chain Management		X	X	X
Game theory and contracts		X	X	X
Systems-dynamics models		X	X	X
Productivity measuring and monitoring		X	X	X
	System Levels of Application			
Tools/Research Areas	**Patient**	**Team**	**Organization**	**Environment**
Financial Engineering and Risk Analysis Tools				
Stochastic analysis			X	X
Value-at-Risk			X	X
Optimization tools for individual decision making		X	X	X
Distributed decision making: market models and agency theory			X	X
Knowledge Discovery in Databases				
Data mining			X	X
Predictive modeling		X	X	X
Neural networks		X	X	X
SYSTEMS-CONTROL TOOLS				
Statistical process control	X	X	X	
Scheduling		X	X	

NOTE: Italics indicate areas with significant research opportunities.

management, and regulation of health care delivery provide few incentives for the use or development of systems-engineering tools. Current reimbursement practices, regulatory frameworks, and the lack of support for research have all discouraged the development, adaptation, and use of systems-engineering tools. Cultural, organizational, and policy-related factors (e.g., regulation, licensing, etc.) have contributed to a rigid division of labor in many areas of health care that has also impeded the widespread use of system tools.

In fact, relatively few health care professionals or administrators are equipped to think analytically about health care delivery as a system or to appreciate the relevance of systems-engineering tools. Even fewer are equipped to work with engineers to apply these tools. The widespread use of systems-engineering tools will require determined efforts on the part of health care providers, the engineering community, state and federal governments, private insurers, large employers, and other stakeholders.

Chapter 3 Recommendations

Recommendation 3-1. Private insurers, large employers, and public payers, including the Federal Center for Medicare and Medicaid Services and state Medicaid programs, should provide more incentives for health care providers to use systems tools to improve the quality of care and the efficiency of care delivery. Reimbursement systems, both private and public, should expand the scope of reimbursement for care episodes or use other bundling techniques (e.g., disease-related groups, severity-adjusted capitation for Medicare Advantage, fixed payment for transplantation, etc.) to encourage the use of systems-engineering tools. Regulatory barriers should also be removed. As a first step, regulatory waivers could be granted for demonstration projects to validate and publicize the utility of systems tools.

Recommendation 3-2 Outreach and dissemination efforts by public- and private-sector organizations that have used or promoted systems-engineering tools in health care delivery (e.g., Veterans Health Administration, Joint Commission on Accreditation of Healthcare Organizations, Agency for Healthcare Research and Quality, Institute for Healthcare Improvement, Leagfrog Group, U.S. Department of Commerce Baldrige National Quality Program, and others) should be expanded, integrated into existing regulatory and accreditation frameworks, and reviewed to determine whether, and if so how, better coordination might make their collective impact stronger.

Recommendation 3-3. The use and diffusion of systems- engineering tools in health care delivery should be promoted by a National Institutes of Health Library of Medicine website that provides patients and clinicians with information about, and access to, systems-engineering tools for health care (a systems-engineering counterpart to the Library of Medicine web-based "clearinghouse" on the status and treatment of diseases and the Agency for Healthcare Research and Quality National Guideline Clearinghouse for evidence-based clinical practice). In addition, federal agencies and private funders should support the development of new curricula, textbooks, instructional software, and other tools to train individual patients and care providers in the use of systems-engineering tools.

Recommendation 3–4. The use of any single systems tool or approach should not be put "on hold" until other tools become available. Some systems tools already have extensive tactical or local applications in health care settings. Information-technology-intensive systems tools, however, are just beginning to be used at higher levels of the health care delivery system. Changes must be approached from many directions, with systems engineering tools that are available now and with new tools developed through research. Successes in other industries clearly show that small steps can yield significant results, even while longer term efforts are being pursued.

Recommendation 3–5. Federal research and mission agencies should significantly increase their support for research to advance the application and utility of systems engineering in health care delivery, including research on new systems tools and the adaptation, implementation, and improvement of existing tools for all levels of health care delivery. Promising areas for research include human-factors engineering, modeling and simulation, enterprise management, knowledge discovery in databases, and financial engineering and risk analysis. Research on the organizational, economic, and policy-related barriers to implementation of these and other systems tools should be an integral part of the larger research agenda.

Information and Communication Technologies for Health Care Delivery

Although information collection, processing, communication, and management are at the heart of health care delivery, and considerable evidence links the use of clinical information/communications technologies to improvements in the quality, safety, and patient-centeredness of care, the health care sector remains woefully underinvested in these technologies (Casalino et al., 2003; DOC, 1999; IOM, 2004c; Littlejohns et al., 2003; Pestonik et al., 1996; Walker et al., 2005; Wang et al., 2003). Factors contributing to this longstanding information/communications technologies deficit include: the atomistic structure of the industry; current payment/reimbursement regimes; the lack of transparency in the market for health care services; weaknesses in health care's managerial culture; the hierarchical structure and rigid division of labor in health professions; and (until very recently) the immaturity of available commercial clinical information/communications systems.

In the past decade, efforts to close the information/ communications technologies gap have focused on the need for a comprehensive national health information infrastructure (NHII), that is, the "set of technologies, standards, applications, systems, values, and laws that support all facets of individual health, health care, and public health" (National Committee on Vital and Health Statistics, 2001). Recent progress toward this goal, including the creation of the Office of the National Coordinator for Health Information Technology (ONCHIT), in the U.S. Department of Health and Human Services, and the release of a 10-year plan to build the NHII, is encouraging (Thompson and Brailer, 2004).

A fully implemented NHII could support applications of information/communications technologies that empower individual patients to assume a much more active, controlling role in their own health care; improve access to timely, effective, and convenient care; improve patient compliance with clinician guidance; enable continuous monitoring of patient conditions by care professionals/care teams; and enable care providers to integrate critical information streams to improve patient-centered care, as well as to analyze, control, and optimize the performance of care teams. The NHII could enable health care organizations to integrate their clinical, administrative, and financial information systems internally, as well as link their systems with those of insurers, vendors, regulatory bodies, and other elements of the extended health care delivery enterprise. The NHII could allow provider organizations to make more extensive use of data/information-intensive systems- engineering tools and facilitate the aggregation and exchange of data among health care organizations, public and private payer organizations, regulatory bodies, and the research community. This data pool could support better regulation and oversight of the health care delivery system, population health surveillance, and the continuing development of the clinical knowledge base.

The NHII could also support another family of emerging technologies based on wireless communications and microelectronic systems with the potential to radically change the structure of the health care delivery system and advance patient-centeredness and quality performance. Wireless integrated microsystems (WIMS) could have an enormous beneficial impact on the quality and cost of health care, especially home health care in the coming decade. Microsystems implemented as wearable and implantable devices connected to clinical information systems through wireless communications could provide diagnostic data and deliver therapeutic agents for the treatment of a variety of chronic conditions, thereby improving the quality of life for senior citizens and chronically ill patients.

Much of the information/communications technology necessary for the realization of NHII exists today and will certainly improve in the decade ahead;

however, there will be many challenges to putting it in place. Interoperability and other data standards and serious privacy and reliability concerns must be addressed, as well as training issues at all levels of the health care system. These and many of the same structural, financial, policy-related (reimbursement schemes, regulation), organizational, and cultural barriers that have impeded the use of systems tools will have to be surmounted to close health care's wide information/communications technologies gap.

Chapter 4 Recommendations

Recommendation 4-1. The committee endorses the recommendations made by the Institute of Medicine Committee on Data Standards for Patient Safety, which called for continued development of health care data standards and a significant increase in the technical and material support provided by the federal government for public-private partnerships in this area.

Recommendation 4-2. The committee endorses the recommendations of the President's Information Technology Advisory Council that call for: (1) application of lessons learned from advances in other fields (e.g., computer infrastructure, privacy issues, and security issues); and (2) increased coordination of federally supported research and development in these areas through the Networking and Information Technology Research and Development Program.

Recommendation 4-3. Research and development in the following areas should be supported:

- human-information/communications technology system interfaces
- voice-recognition systems
- software that improves interoperability and connectivity among systems from different vendors
- systems that spread costs among multiple users
- software dependability in systems critical to health care delivery
- secure, dispersed, multiagent databases that meet the needs of both providers and patients
- measurement of the impact of information/communications systems on the quality and productivity of health care

Recommendation 4-4. The committee applauds the U.S. Department of Health and Human Services 10-year plan for the creation of the National Health Information Infrastructure and the high priority given to the creation of standards for the complex network necessary for communications among highly dispersed providers and patients. To ensure that the emerging National Health Information Infrastructure can support current and next-generation clinical information/ communications systems and the application of systems tools, research should focus immediately on advanced interface standards and protocols and standards-related issues concerning access, security, and the integration of large-scale wireless communications. Special attention should be given to issues related to large-scale integration. Funding for research in all of these areas will be critical to moving forward.

Recommendation 4–5. The committee recommends that public- and private-sector initiatives to reduce or offset current regulatory, accreditation, and reimbursement-related barriers to more extensive use of information/ communications technologies in health care be expanded. These initiatives include efforts to reimburse providers for care episodes or use other bundling techniques (e.g., severity-adjusted capitation; disease-related groups, etc.), public and private support of community-based health information network demonstration projects, the Leapfrog Group's purchaser- mediated rewards to providers that use information/ communications technologies, and others.

Recommendation 4–6. Public- and private-sector support for research on the development of very small, low-power, biocompatible devices will be essential for improving health care delivery

Recommendation 4–7. Engineering research should be focused on defining an architecture capable of incorporating data from microsystems into the wider health care network and developing interface standards and protocols to implement this larger network. Microsystems research should be focused on the following areas:

- integration, packaging, and miniaturization (to sizes consistent with implantation in the body)
- tissue interfaces and biocompatibility for long-term implantation
- interfaces and approaches to noninvasive (wearable) devices for measuring a broad range of physiological parameters
- low-power, embedded computing systems and wireless interfaces consistent with *in vivo* use
- systems that can transform data reliably and accurately into information and information into knowledge as a basis for treatment decisions

A Strategy to Accelerate Change

The committee believes that the actions recommended in this report will accelerate the development, adaptation, implementation, and diffusion of systems-engineering tools and information/communications technologies for health care delivery. However, building the partnership between engineering and health care that will accelerate and sustain progress toward the high-performance, patient-centered health care system envisioned by IOM will require bold, intentional, far-reaching changes in the education, research priorities, and practices of health care, engineering, and management. Building on the experiences of recent large-scale, multidisciplinary, research/education/ technology-transfer initiatives in engineering and the biological sciences, the committee proposes a strategy for building bridges between the fields of engineering, health care, and management to address the major challenges facing the health care delivery system. An environment in which professionals from all three fields could engage in basic and applied research and translate the results of their research and advances both into the practice arena and the classroom, where students from the three disciplines could interact, would be a powerful catalyst for cultural change.

Chapter 5 Recommendations

Recommendation 5-1a. The federal government, in partnership with the private sector, universities, federal laboratories, and state governments, should establish multidisciplinary centers at institutions of higher learning throughout the country capable of bringing together researchers, practitioners, educators, and students from appropriate fields of engineering, health sciences, management, social and behavioral sciences, and other disciplines to address the quality and productivity challenges facing the nation's health care delivery system. To ensure that the centers have a nationwide impact, they should be geographically distributed. The committee estimates that 30 to 50 centers would be necessary to achieve these goals.

Recommendation 5-1b. These multidisciplinary research centers should have a three-fold mission: (1) to conduct basic and applied research on the systems challenges to health care delivery and on the development and use of systems- engineering tools, information/communications technologies, and complementary knowledge from other fields to address them; (2) to demonstrate and diffuse the use of these tools, technologies, and knowledge throughout the health care delivery system (technology transfer); and (3) to educate and train a large cadre of current and future health care, engineering, and management professionals and researchers in the science, practices, and challenges of systems engineering for health care delivery.

Recommendation 5-2. Because funding for the multidisciplinary centers will come from a variety of federal agencies, a lead agency should be identified to bring together representatives of public- and private-sector stakeholders to ensure that funding for the centers is stable and adequate and to develop a strategy for overcoming regulatory, reimbursement- related, and other barriers to the widespread application of systems engineering and information/communications technologies in health care delivery.

Accelerating Cultural Change through Formal and Continuing Education

Making systems-engineering tools, information technologies, and complementary knowledge in business/ management, social sciences, and cognitive sciences available and training individuals to use them will require the commitment and cooperation of engineers, clinicians, and health care managers, as well as changes in their respective professional cultures. The committee believes that these long-term cultural changes must begin in the formative years of professional education. Individuals in all of these professions should have opportunities to participate in learning and research environments in which they can contribute to a new approach to health care delivery. The training and development of health care, engineering, and management professionals who understand the systems challenges facing health care delivery and the value of using systems tools and technologies to address them should be accelerated and intensified.

Recommendation 5-3. Health care providers and educators should ensure that current and future health care professionals have a basic understanding of how systems-engineering tools and information/communications technologies work and their potential benefits. Educators of health professionals should develop curricular materials and programs to train graduate students and practicing professionals in systems approaches to health care delivery and the use of systems tools and information/communications technologies. Accrediting organizations, such as the Liaison Committee on Medical Education and Accreditation Council for Graduate Medical Education, could also require that medical schools and teaching hospitals provide training in the use of systems tools and information/communications technologies. Specialty boards could include training as a requirement for recertification.

Recommendation 5-4. Introducing health care issues into the engineering curriculum will require the cooperation of a broad spectrum of engineering educators. Deans of engineering schools and professional societies should take steps to ensure that the relevance of, and opportunities for, engineering to improve health care are integrated into engineering education at the undergraduate, graduate, and continuing education levels. Engineering educators should involve representatives of the health care delivery sector in the development of cases studies and other instructional materials and career tracks for engineers in the health care sector.

Recommendation 5-5. The typical MBA curriculum requires that students have fundamental skills in the principal functions of an organization—accounting, finance, economics, marketing, operations, information systems, organizational behavior, and strategy. Examples from health care should be used to illustrate fundamentals in each of these areas. Researchers in operations are encouraged to explore applications of systems tools for health care delivery. Quantitative techniques, such as financial engineering, data mining, and game theory, could significantly improve the financial, marketing, and strategic functions of health care organizations, and incorporating examples from health care into the core MBA curriculum would increase the visibility of health care as a career opportunity. Business and related schools should also be encouraged to develop elective courses and executive education courses focused on various aspects of health care delivery. Finally, students should be provided with information about careers in the health care industry.

Recommendation 5–6. Federal mission agencies and private-sector foundations should support the establishment of fellowship programs to educate and train present and future leaders and scholars in health care, engineering, and management in health systems engineering and management. New fellowship programs should build on existing programs, such as the Veterans Administration National Quality Scholars Program (which supports the development of physician/scholars in health care quality improvement), and the Robert Wood Johnson Foundation Health Policy Research and Clinical Scholars Programs (which targets newly minted M.D.s and social science Ph.D. s, to ensure their involvement in health policy research). The new programs should include all relevant fields of engineering and the full spectrum of health professionals.

Call to Action

As important as good analytical tools and information/ communications systems are, they will ultimately fail to transform the system unless all members of the health care provider community participate and actively support their use. Although individuals "on the ground" (i.e., those doing the work) often know best how to improve things, empowering them to participate in changing the system will require that they understand the overall goals and objectives of the system and subsystem in which they work. Based on this understanding, they can contribute substantively to continuous improvements, as well as to radical advances in processes. The communication of the overall system and subsystem goals to individuals and groups at all levels is a crucial task for the management of the organization, and encouraging and recognizing individuals for their contributions to continuous improvements in operations at every level must be a principal operating goal for management.

Overhauling the health care delivery system will not come quickly or easily. Achieving the long-term goal of improving the health care system will require the ingenuity and commitment of leaders in the health care community, including practitioners in all clinical areas, and leaders in engineering. The committee recognizes the immensity of the task ahead and offers a word of encouragement to all members of the engineering and health care communities. If we take up the challenge to help transform the system now, crises can be abated, costs can be reduced, the number of uninsured can be reduced, and all Americans will have access to the quality care they deserve and that we are capable of delivering.

References

Casalino, L., R.R. Gillies, S.M. Shortell, J.A.Schmittdiel, T. Bodenheimer, J.C. Robinson, T. Rundall, N. Oswald, H. Schauffler, and M.C. Wang. 2003. External incentives, information technology, and organized processes to improve health care quality for patients with chronic diseases. Journal of the American Medical Association 289(4): 434–441.

DOC (U.S. Department of Commerce). 1999. The Emerging Digital Economy II: Appendices. Washington, D.C.: DOC.

Feistritzer, N.R., and B.R. Keck. 2000. Perioperative supply chain management. Seminars for Nurse Managers 8(3): 151–157.

Fone, D., S. Hollinghurst, M. Temple, A. Round, N. Lester, A. Weightman, R. Roberts, E. Coyle, G. Bevan and S. Palmer. 2003. Systematic review of the use and value of computer simulation modelling in population health and health care delivery. Journal of Public Health Medicine 25(4): 325–335.

IOM (Institute of Medicine). 2000. To Err Is Human: Building a Safer Health System, edited by L.T. Kohn, J.M. Corrigan, and M.S. Donaldson. Washington, D.C.: National Academies Press.

IOM. 2001. Crossing the Quality Chasm: A New Health System for the 21st Century. Washington, D.C.: National Academies Press.

IOM. 2004a. Insuring America's Health: Principles and Recommendations. Washington, D.C.: National Academies Press.

IOM. 2004b. Keeping Patients Safe: Transforming the Work Environment of Nurses. Washington, D.C.: National Academies Press.

IOM. 2004c. Patient Safety: Achieving a New Standard of Care. Washington, D.C.: National Academies Press.

Leatherman, S., D. Berwick, D. Iles, L.S. Lewin, F. Davidoff, T. Nolan, and M. Bisognano. 2003. The business case for quality: case studies and an analysis. Health Affairs 22(2): 17–30.

Littlejohns, P., J.C. Wyatt, and L. Garvican. 2003. Evaluating computerized health information systems: hard lessons to be learnt. British Medical Journal 326(7394): 860–863.

McGlynn, E.A., S.M. Asch, J. Adams, J. Keesey, J. Hicks, A. DeCristofaro, and E.A. Kerr. 2003. The quality of health care delivered to adults in the United States. New England Journal of Medicine 348(26): 2635–2645.

Murray, M., and D.M. Berwick. 2003. Advanced access: reducing waiting and delays in primary care. Journal of the American Medical Association 289(8): 1035–1040.

National Committee on Vital and Health Statistics. 2001. Information for Health: A Strategy for Building the National Health Information Infrastructure. Available online at: http://ncvhs.hhs.gov/nhiilayo.pdf.

Pestonik, S.L., D.C. Classen, R.S.Evans, and J.P. Burke. 1996. Implementing antibiotic practice guidelines through computer-assisted decision support: clinical and financial outcomes. Annals of Internal Medicine 124(10): 884–890.

Starfield, B. 2000. Is U.S. health really the best in the world? Journal of the American Medical Association 284(4): 483–85.

Thompson, T.G., and D.J. Brailer. 2004. The Decade of Health Information Technology: Delivering Consumer-centric and Information-Rich Health Care: Framework for Strategic Action. Washington, D.C.: U.S. Department of Health and Human Services.

Walker, J., E. Pan, D. Johnston, J. Adler-Milstein, D.W. Bates, and B. Middleton. 2005. The value of health care information exchange and interoperability. Health Affairs Web Exclusive (Jan.19): W5-10—W5-18.

Wang, S.J., B. Middleton, L.A. Prosser, C.G. Bardon, C.D. Spurr, P.J. Carchidi, A.F. Kittler, R.C. Goldszer, D.G. Fairchild, A.J. Sussman, G.J. Kuperman and D.W. Bates. 2003. A cost-benefit analysis of electronic medical records in primary care. American Journal of Medicine 114(5): 397—403.

Questions to Consider

- A strong point in the executive summary states, "relatively little technical talent or material resources have been devoted to improving or optimizing the operations or measuring the quality and productivity of the overall U.S. health care system." Has this changed? Explore other professions like transportation infrastructure, and identify the federal dollars devoted (US DOT) and the number of civil engineers involved in the support of the seaports, railways, and highways. How does this compare to health care (focus on the involvement of non-clinicians)? Clearly, health care expenditures are a large portion of the gross domestic product (about 20%), but how do these costs relate to the delivery of safe and effective patient care?
- When the report was published, many clinical engineers were thrilled to see the phrase "Engineering-Health Care Partnership" in the Executive Summary. Many felt this was the turning point in not only the quality of the partnership but the quantity as well. Applying system engineering principles seemed to be the key to integration of the two disciplines. When you think about health care delivery 15 years later, have successes been made? Think about the components of systems engineering described, "statistical process controls, queuing theory, quality function deployment, failure-mode effects analysis, modeling and simulation, and human factors engineering." Have these tools (useful by clinicians) been adapted to health care *technology* in addition to the practice of medicine?
- Certainly, the evolution of the electronic medical record does support improvement in the IT sector of health care. However, look at recommendation 4-3. Are any of these areas supported today? Is interoperability and connectivity between devices a reality?
- Explore recommendation 5-4. Perhaps this is the most shocking of the recommendations because it is directly associated with clinical engineering. Only a tiny fraction of the biomedical engineering programs in the country have a clinical focus or are associated with the "health care delivery sector." How can this be improved? How can a better understanding of the use of medical technology as part of patient care supplant the extensive and financially beneficial fundamental biomedical research that is most pervasive in biomedical engineering academic programs?

Article 5: Managing The Lifecycle of Medical Equipment, Tropical Health and Education Trust (THET), London (2015)

About this Guide

Health Partnerships working in low-resource settings frequently encounter challenges relating to medical equipment that can influence the success of their projects. These challenges include a lack of functioning equipment, and other aspects of what is called 'Healthcare Technology Management (HTM)'. HTM concerns the management of the medical equipment life cycle; from planning to purchase, installation, operation all the way through decommissioning and disposal.

This resource serves as a companion to the **Making it Work** toolkit, published by THET in 2013 and offers an overview of the steps of the equipment life cycle and ways for partnerships to integrate these considerations into their projects.

This resource identifies 'Assumptions'; expectations which might be valid for high-resource settings but which are not necessarily valid for low-and middle-income countries (LMICs). These are linked to 'Mitigations'; potential ways to prevent setbacks and to improve the progress of the project and the quality of healthcare in the LMIC. Some of the mitigations need the support of a technical expert, but many can be done without additional resources.

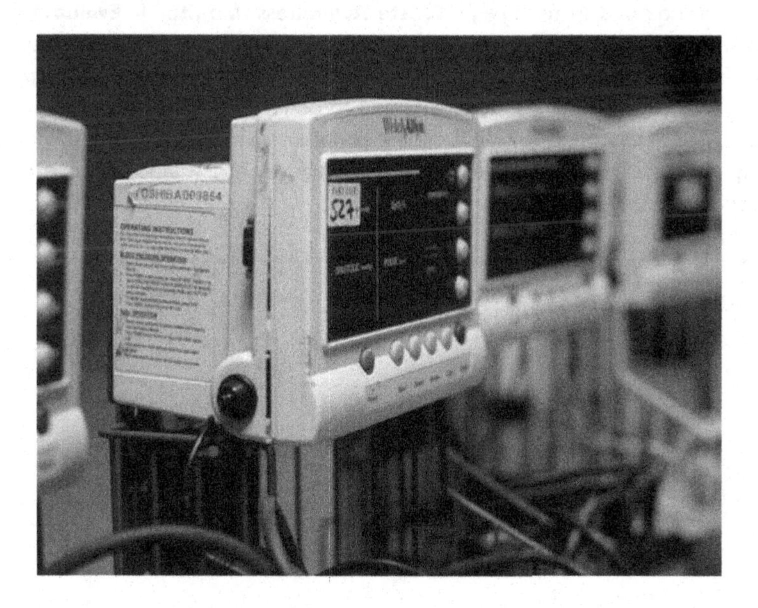

The Equipment Life Cycle

This resource follows the Equipment Life Cycle as it is often used in Healthcare Technology Management (HTM). The cycle is divided in 4 phases and 9 topics.

The first phase 'Planning' consists of **Planning and Assessment** of the needs in the healthcare facility appropriate to its environment, the equipment users and patients, and **Budget & Financing** in which the appropriate budgets are created and estimated for purchase and the 'cost of ownership'.

The second phase 'Purchase' contains **Assessment and Selection**, covering how to decide which equipment meets the needs identified earlier. Specifications are written and in **Procurement & Logistics** a tender is written, a less complicated purchase is done or a donation is agreed upon. The responsibilities and practicalities around logistics are prepared and executed. In **Installation & Commissioning** after the equipment has arrived in the healthcare facility and should be unpacked, installed, and commissioned.

After these two phases of preparation the third phase is the actual 'Lifetime'. Starting with the training of users and maintainers in **Skill Development & Training**, the daily **Operation & Safety** for and by users, and **Maintenance & Repair** mostly done by the Biomedical Equipment Professionals.

The last phase 'End of Life' is about **Decommissioning & Disposal** of medical equipment.

As indicated in the image, **Create Awareness, Monitor & Evaluate** are constant throughout the life cycle. Creating awareness with all participants, whether they are users, maintainers, administrators or politicians, is of great importance to improve systems and add to better biomedical and healthcare practices. Monitoring and evaluating contributes to keeping track of the equipment lifecycle, and creates opportunities to review and improve processes and share successes and learning.

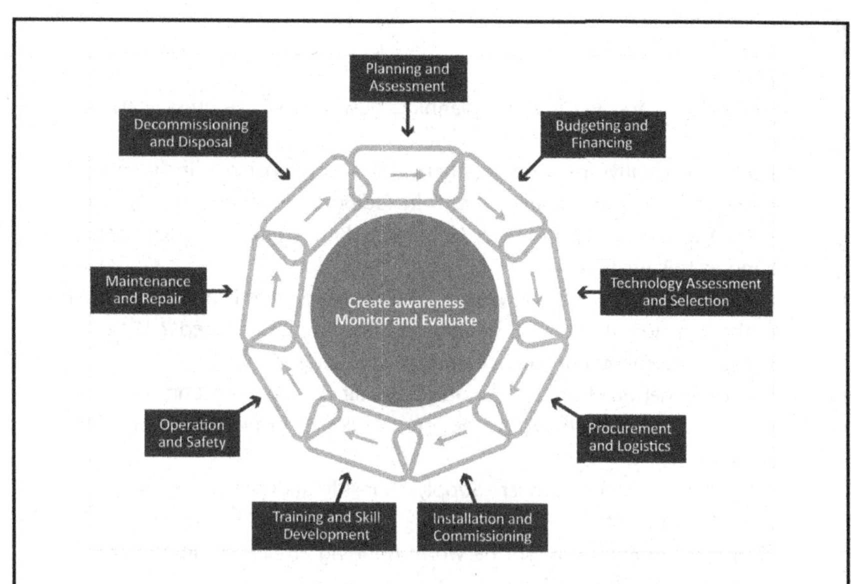

Always Involve Local Technical Staff!

Throughout this resource this symbol will indicate the suggested involvement of a Biomedical Engineer (BME) from your UK trust. The added value of a BME in your team is well illustrated in case study 7 of the Donations Toolkit on p.71. However the involvement of local technical staff in the destination institution should always come first. When no local technical staff are present, it is worth looking for a local contractor.

Phase 1: Planning & Assessment

The assumptions and mitigations described below apply to both planning donations and locally purchased equipment. For detailed information on Medical Equipment Donations, see the **Donations Toolkit** Chapter 1 and 2.

Assumptions

- Safe and stable electrical supply and clean running water is always available, as are medical gases
- Supporting departments function well and deliver quality controlled outputs e.g. sterilisation and laundry departments
- Data is available on which to base decisions on equipment purchases, like user and environmental data, appropriateness to setting, information from this and other hospitals
- There is consensus on and prioritisation of what is required. Users, maintainers, financers and managers give their input and requirements are written with everyone's agreement
- Long term plan (+budget) is in place for equipment purchases

Mitigations

Consider all the following when planning how you will address Planning & Assessment.

- Do a collaborative needs-assessment (UK and DC partner, including technical staff, users and management) including an inventory check, or creation of an inventory. Consider bringing a BME from your UK hospital to support this process
- Do an infrastructure check; what is available and what is working properly. Is there a non-electrical alternative for the identified needs? Work with robust equipment, plan a back-up (e.g. a generator)
- Do additional purchases (e.g. water filter, air-conditioning unit) and attach protective equipment like a stabiliser or UPS to protect equipment from surges
- See Understanding Power Supply Considerations on p. 44 of the Donations Toolkit
- For bigger projects it might be worth bringing an electrician and plumber to site to make basic infrastructure improvements
- Check if supporting departments are functional and effective and take action if necessary
- Coordinate with other agencies, the government and other hospitals. Learn from others by finding out e.g. which organisations work in the same or similar hospital. Have they purchased equipment? How is equipment normally planned and purchased? Often the Ministry of Health (MoH) is in charge of centralised procurement and it is important to understand the dynamics between the parties.

For the role of the Ministry of Health in medical equipment management see p.24 of the donations toolkit

- Consider both the patient journey and all factors of service delivery when creating specifications
- Be prepared for reactive rather than planned purchases. Create awareness of expected lifetimes and long term planning.

Although reactive purchases are often related to limited financial resources, it is important to create awareness on how working equipment is a source of income. An Equipment Development Plan can be found in Ziken's guide 2 Chapter 7.1. This information should be shared with hospital directors, financial managers, procurement officers, users and maintainers

Reactive Vs. Planned Purchase

Medical equipment is valuable and the purchase/tender process takes time. In the UK equipment is mostly replaced before the old equipment is permanently out of service. The Biomedical Technicians know when equipment reaches the end of their profitable life (when the cost of repair and downtime become too high), the users (doctors and nurses, but maybe also cleaning staff) know when equipment lacks functionality or speed. Before a tender process is initiated an internal process takes place in which the hospital prioritises the needs for the coming year(s). The users/departments make a request for a new piece of equipment, the technicians support the proposal with technical background and the financial department prioritises the request, which is then approved by the hospital director/direction. Normally not all requests are accepted due to limited budgets, but when the same request is proposed e.g. two subsequent years, the need is clear. This is called a planned purchase.

In developing countries purchases (or often donations) are done centrally by the MoH. This can be a random process in which users and technicians not always have a say. Purchases are often done after equipment has been out of service for a long time. For example: a district hospital's X-ray is out of service. It takes 6 months before a proper diagnosis is made (no service engineer in the country). It appears the tube is broken, and replacing a tube is a huge investment. The machine is already over 20 years old and it is decided it should be replaced. A request from the hospital to the MoH for a tender is done (in writing) and 3 months later the MoH decided to start a tender procedure. It is to be expected that it takes at least 1 year to execute the tender procedures, accept a bid, place the order, receive and install the equipment. The hospital in this example has to refer its patient for x-rays for almost 2 years before having solved the issue. Referring patients is inconvenient and leads to a loss in revenue.

See Phase 4: Procurement for an example of centralised procurement in the UK.

"An early intervention at Connaught Hospital was a full inventory of all hospital equipment. We were therefore able to work with hospital staff to redistribute existing equipment (much of which was needed but unused) and identify critical gaps."

DR OLIVER JOHNSON,

King's Health Partners, Programme director King's Sierra Leone Partnerships

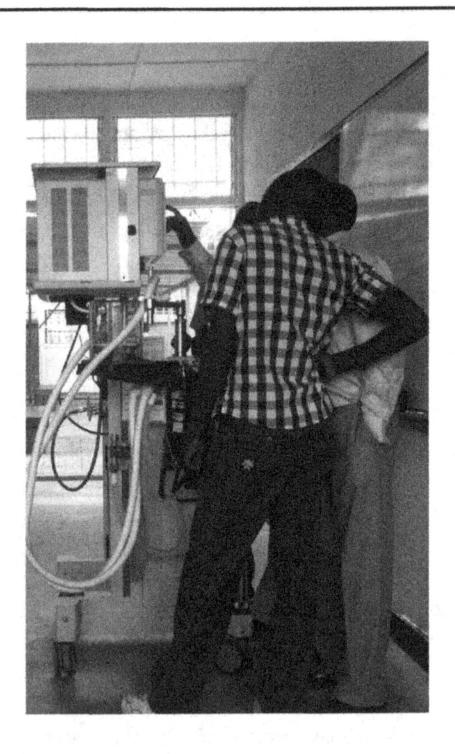

Phase 2: Budgeting & Financing

Assumptions

- Hidden costs are covered and planned for, e.g. maintenance, HR, training, consumables, replacements
- Financial management and rules are understandable, available and applied
- Budget is existent, usable and realistic & implies responsibility/planning for the future
- Spare parts and consumables are available for reasonable prices

Mitigations

- Share the hippo model (see below). Create awareness and encourage budgets to be created for the equipment lifetime (Cost of ownership estimated by 10% of purchase cost /yr)
- Describe an equipment situation to show that maintenance makes economic sense
- Insist on transparent processes, for example by proposing the use of the long-term Equipment Development Plan and Core Equipment Expenditure Plan as described in chapter 7.1 and 7.3 of Ziken's Guide 2
- Clarify responsibilities & cost allocation, encourage flexibility on allocations

- Prioritise needs and link to available budgets to create a feasible plan
- Make use of local/historical knowledge & ownership e.g. local purchasing
- Research the availability of spare parts, consumables, and maintenance services. Try to avoid importing parts yourself; the local system should be encouraged and local economies stimulated.
- Learn from the BMEs in your UK hospital

More information on budgeting for medical equipment can be found in Ziken's Guide 2. Guide 6 covers the financing of Medical Equipment.

In Focus

The partnership between Guy's & St Thomas' NHS Foundation Trust, Arthur Davison Children's Hospital and Ndola Central Hospital in Zambia was set up in 2009, focusing on improvement of biomedical services in those two hospitals reaching out to other biomedical professionals in the Copperbelt region as well. The lack of spare parts has been a challenge and focus for this project. The Zambian government has procurement regulations that do not allow public hospitals to order parts from outside the country (for example online). The few Zambian medical equipment suppliers present in the country triple or quadruple prices and are in somewhat of monopoly position. The lack of competition and market control allows them to maintain this position. A potential solution that is currently being explored is to ask a local hardware store to order online and have a small commission. Often it is not necessary to be a formal agent to be able to order spare parts. In the meantime cases should be reported to the Ministry of Health to raise awareness and fight for improvement of the current situation and regulations.

"We use an ultrasound to identify liver disease/cancer in patients, which is non-invasive, quick and acceptable to patients. This machine often broke down, due to lack of care/maintenance on the local site. Also, the high temperatures often contributed to the machine malfunction. Without the machine, accuracy of patient diagnosis was limited and it slowed the project down. The latter was due to the need to undertake a biopsy to diagnose disease. This is invasive, disliked by patients and requires a skilled surgeon, requiring additional resources to obtain confirmation of those patients with liver disease. Lack of this data would limit the data and effectiveness of the project."

"We procured additional (backup) ultrasound machines to cover for breakdown and had medics experienced in using/caring for the machine spend short intensive periods in Africa diagnosing the patients. We also paid for regular machine maintenance/service to keep the machines active. Both solutions allowed diagnosis of patients and sufficient data for the project outcomes."

DR D GARSIDE

Imperial College London - Gambia partnership, Project Manager

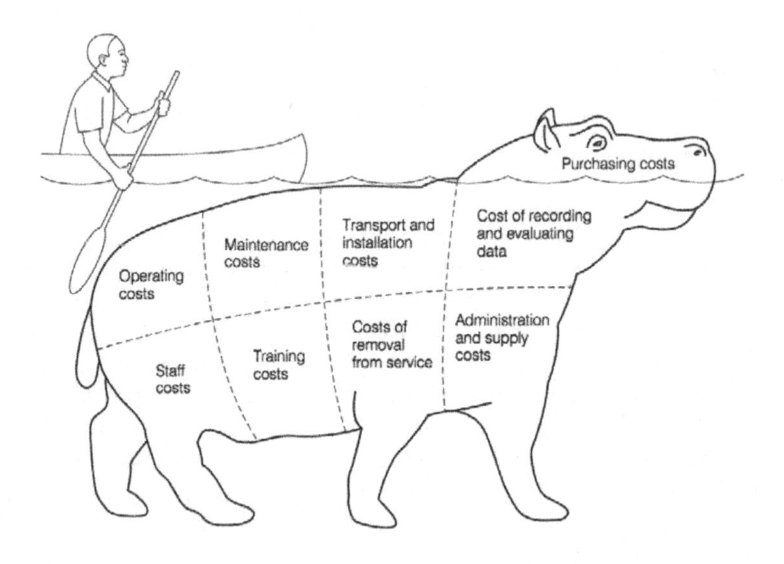

The hippo model

When purchasing (medical) equipment, care providers should budget and plan for all cost hidden under water level; Purchasing costs cover only a minor part of the total cost of ownership.

The Hippo model is an alternative way of depicting the iceberg, which can be found on p. 10 of the Donations Toolkit p. 10

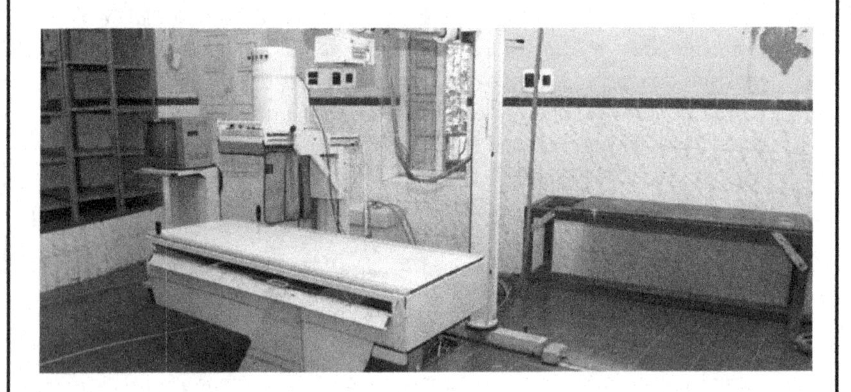

Phase 3: Technology Assessment & Selection

Assumptions

- All equipment is available to be purchased and within reach & you can trial it before purchase
- Users know how to use equipment and are systematically educated
- What is advertised (equipment + service) is available
- You can trust the market to deliver equipment of good quality and safety
- Manufacturers or agreed agents are locally present
- Qualified and trained technicians are locally present
- Local spare parts stock is present
- Honesty & ethics of manufacturer are strong
- Consumables and spare parts continue to be available throughout the lifetime of the equipment
- Equipment fits the purpose and is appropriate to setting
- "Household name" or "well-known brand" companies operate in the same manner in an emerging economy as they do in the UK

Mitigations

- Pilot the equipment, visit the agents or vendors, share information and experiences with other parties/hospitals (try available equipment in other hospitals, and look for existence of national standards (if not, use European standards), verify the reliability of vendors)
- Establish training needs, including basic awareness of safety and equipment care
- Check what local vendors can deliver on, which timescale and what kind of service they offer. Meet the vendors, check their facilities, and their reputation.
- Stick to FDA and/or CE medical marked equipment. Do not fall for cheap options. Check if the Ministry of Health has adopted standards and regulations on medical equipment
- Check which vendors are present in the country and if they are recognised by the manufacturers. Think of service support as well, check the presence of licensed service engineers.
- Internationally recognised manufacturers do not gamble with ethics and honesty. They avoid risks to their reputation. However, it is advisable to ask around for references. Check if the Ministry of Health keeps a black list.
- Check the availability (and price! !) of spare parts and consumables beforehand. Consider re-usable accessories for remote areas but bear in mind that this only works if sterilisation is done properly. Consider simpler equipment to avoid the use of expensive spare parts. Use whole-life cycle

costing, and write a tender for spare parts or ask for price guarantees for 3 years.

- Check specifications on appropriateness to setting and, during assessment, include local productions or nonprofit equipment that is developed for low-resource settings. Also consider standardising the equipment; if all public hospitals use the same brand equipment, it might be advisable to purchase the same. This helps to secure access to service and parts.
- Learn from the BMEs in your UK hospital

See p.25 of the Donations Toolkit for 'Asking the right questions' to understand whether the equipment is appropriate to setting

"We standardised our BP, pulse, temp and sats monitors on the wards, and bought the most simple to use and maintain."

DR BIPLAB NANDI

Queen Elizabeth Central hospital Blantyre, Malawi & Great Ormond Street Hospital London, developing country lead

In Focus

Rwanda has introduced law saying that no second hand equipment can be brought into the country. For donations and refurbished equipment this can mean that equipment is not cleared and sent back to its origin at a cost to the sender. Although secondhand high-end equipment might be more appropriate to the setting (safe and reliable) than new Chinese equipment, often these rules are strict and without exceptions.

When purchasing equipment there are roughly 4 options:

1. New equipment from big manufacturer
   - Plus + good quality
   - Plus + access to service, spare parts and consumables
   - Minus - expensive in purchase
   - Minus - difficult and expensive to maintain/repair
   - Minus - more functionality than necessary
2. Second-hand or manufacturer refurbished equipment from big manufacturer
   - Plus + less complicated in use
   - Plus + attractive price/quality
   - Plus+ Refurbished equipment might come with a guarantee for availability of spare parts and consumables
   - Minus - not as desirable as new equipment (wanting the 'gold standard)

3. Equipment produced for low-resource settings (often start-ups or NGOs)
   - Plus + appropriate to setting (functionally and technically)
   - Plus + not expensivec
   - Minus - unsure if the company will last (availability of spare parts)
   - Minus - not as desirable as new equipment (wanting the 'gold standard)

   The Donations Toolkit mentions several of these initiatives on p. 41 "Supplying Appropriate Technologies Designed for Low-Resource Settings" and p. 81 for contact details
4. New equipment of inferior quality mostly produced in Asia
   - Plus + not expensive
   - Plus + fast delivery
   - Minus - no quality guarantees (CE/FDA)
   - Minus - access to service/spare parts
   - Minus - short life time
   - Minus - higher level of break downs

Phase 4: Procurement & Logistics

Assumptions

- Tender procedures are well known and respected
- Logistics are costed, including customs and transport (effective/reliable/timely and safe) from port to hospital
- Supplier is honest & efficient
- Specifications are relevant & appropriate
- Company honours warranty
- User knows warranty is there, and can use the information
- In case of accidents there is insurance in place

Mitigations

- Follow local rules e.g. customs and use local experience. Often the ministry of health centrally procures medical equipment and knowledge of tender procedures and logistics are available there

  The process of Clearing Customs is well described in Chapter 6 of the Donation Toolkit and can be found on p. 61
- Include transport in specifications. Delivery in port/airport or in the hospital? Best to include transport until the exact place of installation.
- Check if the space in the hospital is available and appropriate. Go and look.
- If supplier does not do clearance and local transport, prepare a transport plan and ensure reliable carriers, who take ownership for each leg of the journey. Include worst-case scenarios.

  More information can be found in Donations Toolkit Chapter 5, p. 51–59
- Access standard specifications (WHO. Nepal)
- Get references on reliable partners - use consumer power
- Make use of a pre-purchase demo or loan
- Make sure user knows exact warranty conditions and has access to service provider
- User involvement in every stage of the procurement process
- Verify if all stages of the transport are insured and under which conditions.
- Learn from the BMEs in your UK hospital

  See p.52/53 of the toolkit

For more information on logistics see the Donations Toolkit Chapter 5 and 6, and Ziken's Guide 3 accurately describes all elements of Procurement and Commissioning

Centralised Procurement in the UK

In the UK Hospitals procure their own equipment, but often use joint supply agencies ('consortium') to process. That route uses some bulk discount, and there is an 'NHS catalogue' of approved products and prices. So it is a sort of prequalified system, but hospitals are free to act on their own.

In Focus

In the Comoro Islands, technicians were receiving a container with an X-ray in the port of Anjouan. When they opened the container, the forklift was struggling to get the crate out of the container and the technicians assumed the wood had warped, and consequently it was jammed in the container. After transporting the equipment to the hospital the technicians started installing the equipment and found out it was broken. Although the crate was not visibly damaged, apparently an impact from outside had bent the container wall, crushing the equipment inside. No proof was present that the damage was caused during transportation and insurance didn't want to take responsibility. Therefore, ALWAYS check all packaging before opening, and take photos in case of abnormalities. And only remove crates when they have arrived at the final destination. Crates also protect during local transport. Report to supplier, insurance and transporter within 24hours in writing, adding photos.

Phase 5: Installation & Commissioning

Assumptions

- Facilities exist and are appropriate, e.g. space to store the equipment, doors big enough for equipment entry, floors strong enough, water and power supplies are available
- Room preparation needs assessment has been done; everyone knows what needs to be done
- Room preparations are done
- Someone will receive it at site, supervise and sign off the installation
- The equipment is delivered and installed by the supplier
- Test equipment and skilled technicians are present to perform functional and safety tests
- Financial penalties for delays are well communicated and understood by all parties

Mitigations

- Perform a Needs Assessment, create plan for room preparation
  More information Pre-installation work can be found in Ziken's Guide 3 p.200 and estimation of pre-installation cost in Zi ken's guide 2 pill
- Cross-department communications and agreement on who is responsible for which part of the installation and commissioning
- Follow up on room preparation plan, check well in advance
- Plan for user approval on delivery (no damage, is it well installed, is it functioning properly? — standard forms available)
- Let the vendor's service engineer open the boxes, let it be supervised by the hospital's technician
- Makes sure this is included in the tender document or purchase agreement
- Ideally the supplier performs installation and tests under supervision of the hospital technician (directly training the technician). Often test equipment is not available and if available the technician does not always know how to use it. Providing the technician with test equipment and following the Acceptance log sheet helps the partnership to be guided through all the possible checks, but bringing a UK Biomedical Engineer with test

equipment for a release test visit (and training) might be the most feasible solution

Chapter 4 of the Donations Toolkit for more information on verifying the quality and safety of equipment, p48 onwards.

- Financial penalties and insurance clarified on delays, damage and malfunctioning equipment
- Warranty commences and payment made only after successful installation
- End users are aware of warranty conditions. Confirm in writing that the supplier will honour the warranty if purchased in-country
- For smaller items that do not need installation the reception process should be well planned as well. The content of the boxes should be checked against the packing list and the content should be checked on completeness and functionality. In case of discrepancies the supplier should be contacted directly.

For more information on receiving equipment: Donations Toolkit Chapter 6 and Ziken's Guide 3

Acceptance log-sheets guide technicians through the procedure of receiving, testing and installing equipment. Such a sheet is an extensive document of about 10 pages and includes all steps to be undertaken, such as technical tests, execution of training of personnel, presence of manuals, consumables and spare-parts. An example of an acceptance log-sheet can be found in Ziken's Guide 3 p332.

In Focus

It happens that hospitals are not aware of the arrival of medical equipment. Often these are donations, agreed upon by a certain doctor or administrator or the central government deciding equipment should go to this place. Many hospitals in developing countries have a lack of space. When a piece of equipment arrives without notice, it can happen that this equipment sits outside until space is created. This can take a while, with a lack of ownership and awareness, a rainy season and a dry season and the equipment is rusted and rotten without having been used at all.

Testing Equipment

Mulago National Referral Hospital in Uganda has not had access to test equipment for many years. Once the devices are fixed, the technicians have to rely on the users to tell whether they are functioning normally. Recently, new test equipment have been donated and the hospital technicians are slowly getting adapted to their use. Oxygen concentration test device is missing yet the hospital produces its own oxygen. Volunteers are routinely asked to bring some of these tester around to test for the concentration.

Phase 6: Training & Skills Development

Assumptions

- People are used to working with technology
- People have had full medical training and participate in/have access to continuous professional education
- Training is seen as good for skills and prospects both at management level and working on the floor
- Training is included in a tender and executed by the supplier
- Training takes place between installation and taking the equipment into service
- Training is repeated if needed

Mitigations

Refer to Donations Toolkit Chapter 7 - putting the equipment into service, p.67 training of users and maintainers

Consider all the following when planning how you will address training and skills needs.

- Include training in tender specifications (describe needs), and specify who should be trained for how many days with what outcomes
- Cover essential safety and care before putting equipment into service for both maintainers and users
- Begin by doing an assessment of current knowledge, both for users as technicians. Consider bringing a UK BME to identify the needs
- Create training that fits the local needs. The materials and examples used in the training should resemble reality
- Ensure training includes assessment of individual competencies
- Build motivation for the future, explain how training can increase status and respect
- Identify champions, train the trainer, to guarantee continuation of training for new staff and repetition for current staff
- Repeat user training every 6 months, for changing staff. It is possible to include follow-up training in a tender, e.g. 50 hours of training in the following 2 years.
- Explain to management the value of training
- Give the BMET the responsibility for user training; let him/her join the vendor's training. Collaborate with Head of Departments for planning and content.
- Award trained people with a certificate

Ziken's Guide 3 covers initial equipment training and Guide 4 covers user training

Local Champions

In every department, team or professional group you can find champions. Potential champions are those who pay serious attention to the subject, who ask the most questions and who want to talk to the teacher at the end of the class. When you are looking for sustainability of your training, you should look for people who can perform your training in the future. Identify a potential champion and help him/her to get a champion status by providing extra time with him/her, asking him/her to share or take over your class, or even taking him/her to lunch: rewarding their effort and creating a status that will support them to perform training in the future.

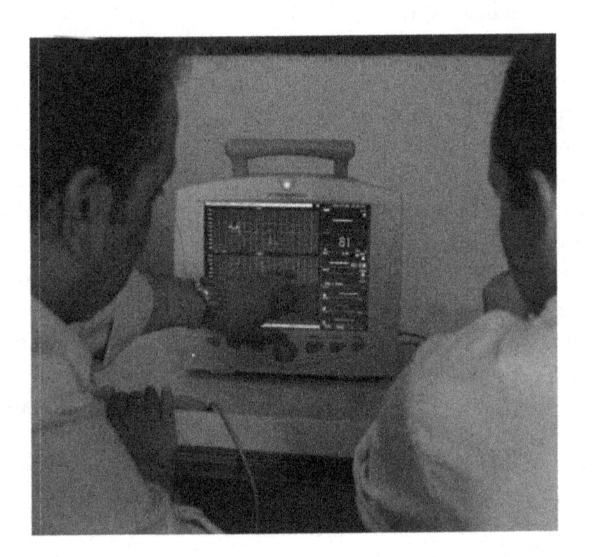

User Training

Biomedical Equipment professionals are often not well respected in the hospital, due to the invisibility of their work. By making the BMET responsible for executing regular user training(s) he/she has the opportunity to make him/herself visible and to spend some time on explaining his role in the healthcare system. This only works with support from the head of departments, the head of nursing and administrators.

> *"We try to teach the importance of maintaining equipment when we are there, and produce guidance on maintenance on simple documents. We always take one team member now who has better understanding of the equipment that we have introduced, such as the oxygen concentrators, and spends time with potential maintenance champions at the hospital."*

FRANKIE DORMON

Medical Lead in Poole Africa

"We saw student nurses and midwives trained in a lovely new college, with excellent equipment then going out to clinical areas and experiencing little equipment and what there is being of poor quality or not working. This is demotivating for staff and unhelpful for patients. There are sparse supplies of oxygen for example and so nurses in the special care baby unit have to decide which babies get it and which don't."

SANDRA PAICE

Juba link Isle of Wight, Nursing and midwifery advisor

Coincidentally, the first cohort of the Rwandan BMET training in Kigali had the opportunity to spend a day with a representative of Zeiss, training the technicians on the working principles and basic maintenance of microscopes. The students insisted on receiving a Certificate, which was created, printed, and signed on the spot. The value of training is not only in increasing your knowledge, but also in having proof of the trainings you've participated in.

Phase 7: Operation & Safety

Assumptions

- Training is followed, assimilated, practiced, and knowledge shared
- Governance & training of trainers is in place
- There is a safety culture and personal protection is available
- The hospital is clean and hygiene is highly respected
- Patient Safety comes first, protocols exist, are used and respected
- People will say when they need training
- Errors are reported and followed up
- Equipment present is working
- Single use consumables are disposed after use.
- The sterilisation service delivers clean and sterile devices.

Mitigations

For more information on using and maintaining equipment see the Donations Toolkit Chapter 7, p70

- Plan for refresher training, Train the trainer, BMET to remind heads of departments to organise trainings. Encourage briefings and debriefings for exchange of knowledge
- Do safety checks eg: every 3 months, train on awareness and safety practices. Check personal protection is available (e.g. gloves, face masks but also radiation protection items like aprons)
- Organise training on sterility and hygiene. Check what products are used to clean. Do not only focus on the cleaning staff. Hygiene is a basic skill for everyone working in a healthcare setting.
- Introduce good practice protocols and train the staff how to use them
- Encourage staff to identify their needs with head of departments and other leaders
- Create awareness around errors and how we can learn from them. Avoid guilt and blame culture. Introduce anonymous reporting to be able to track errors and create an investigation structure
- Check if equipment is operational. Bring or report malfunctioning equipment to the BMET department. Remove faulty equipment from the workspace.
- Consider reusable consumables and not single use. Verify the quality of the sterilisation equipment
- Teach technicians or sterilisation staff how to clean, disinfect and sterilise devices, and to verify whether autoclaves are working (measure pressure and Temperature cycles)

Ziken's Guide 4 describes all elements of daily operation and safety of medical equipment

For more information on logistics see the Donations Toolkit Chapter 5 and 6, and Ziken's Guide 3 accurately describes all elements of Procurement and Commissioning

Protocols

Medical guidelines or protocols are not always common in developing countries. Introducing best practices guidelines in trainings and distributing them/sticking them to wall helps staff to work consistently. The WHO has developed some useful tools as well, like the surgical safety checklist.

In Focus

Mariette Jungblut, an expert of sterile medical devices from the Netherlands, was teaching about sterility and hygiene in a South-African nursing college when she came across cleaners that disinfect the entire hospital with chlorine.

Chlorine is very aggressive and not suitable to clean medical equipment or e.g. mattresses with. Hospital mattresses are supposed to be watertight, to prevent body fluids to enter the foam, but by using chlorine, the cover becomes porous and the mattress far from hygienic. Her advice: stick to cleaning with soap and warm water. Use chlorine only on floors, walls or sanitary if it is soiled with body fluid. Never use chlorine to disinfect medical instruments, because corrosion will destroy your instruments. Good hygiene is cleaning with soap and warm water.

Phase 8: Maintenance & Repair

Assumptions

- The environment in which the equipment is used is stable and known (24h/24h).
- Maintenance culture exists and is respected by the technicians, users and other staff
- Technical staff present, trained and know how to maintain and repair the equipment
- Technical staff respected
- Preventive maintenance (PM) schedules exist and PM is performed regularly
- Technicians have access to an equipped workshop
- Technicians have access to spare parts, on stock in the hospital or ordered in and spare parts are delivered within 24 hours if necessary
- Technicians have access to digital or paper service and user manuals
- Technicians have access to and know how to use test equipment to calibrate and test medical equipment
- Users know how to use and take care of the equlpment

Mitigations

For more information see the Donations Toolkit chapter 7; using and maintaining the equipment p. 70

- Prepare for environmental challenges, e.g. humidity, dust and heat
- In case of a lack of technical staff, see if there is a way to create contractual obligations to support maintenance
- identify the technical staff, get an idea of their skills and knowledge and encourage/organise training
- Help technicians to structure their ways of working and spread these principles in the hospital (e.g. users understand what to do with broken equipment). Help to create visibility and encourage technicians to keep track of their work and successes, to be able to report to the hospital director. Consider inviting a UK BME to your team to cover this work

- See if the technicians make use of Planned Preventive Maintenance Schedules and if not, create them for your most crucial equipment. Instructions can be found in the Service manuals
- Check what space and tools the technicians have to perform maintenance and repair. In case of insufficient infrastructure, it is worth the effort to create an inventory, identify the needs and write to the director/MoH.

See the Donations Toolkit p.45 for more information on sourcing biomedical engineering tools and test equipment

- Check that supply chains for service support exist. Access to spare parts is one of the biggest challenges for biomedical technicians in low-resource settings. Estimate in advance spare parts and consumables needs, and discuss budget needs and supply chain
- Often Medical Equipment in developing countries is donated and manuals are not present. Manufacturers are protective of their manuals and these are normally not easy to find online. See box below for available resources
- Consider bringing a UK BME with test equipment to check crucial equipment for safety and quality
- Often equipment failure is caused by user errors. Train the users to properly operate the equipment but also to take care of the equipment. Most of the weekly preventive maintenance can be performed by the users (e.g. nurses can clean filters)
- Label all tools and test equipment and inform management about them. Nominate a person to be in charge of them and have others to sign them out and upon return so that equipment are not easily lost

See Ziken's Guide 5 for more information on Maintenance Management of Medical Equipment.

Service manuals are often missing in developing countries' hospitals, and it's difficult to find manuals online. However there are several resources where we can find manuals:

The manufacturer, the UK trust biomedical workshop, **Frank's Hospital Workshop** the INFRATECH mailing list and manuals collected by the French NGO **Humatem**.

Also see the Donations Toolkit p. 44 "Getting the right service manuals"

Phase 9: Decommissioning & Disposal

Assumptions

- Disposal channels are available for when equipment reaches the end of its life
- When disposing equipment the environment is considered
- There are clear regulations on waste disposal
- Companies that buy old equipment exist

- Decommissioning regulations exist, e.g. erasing of patient data and decontamination and the technicians know how to do this
- When purchasing new equipment the supplier may take responsibility for the equipment that is being disposed

Mitigations

- Create awareness and share best practices on disposal from the UK
- Awareness-raising, explain the environmental impact
- Encourage hospitals to create disposal routes and raise awareness on Ministry level
- Teach technicians how to decommission, e.g. decontaminate and erase patient data
- Include disassembly and disposal of equipment in the tender specifications, consider if that is acceptable for the owner (the hos pita l/M oH might see a value – auction to scrap buyers. Try to convince that cleaning up is a more suitable solution than keeping a junkyard)

Monitoring, Evaluation & Learning

Make sure that monitoring and evaluation is on-going process by establishing the systems that you will use to gather, manage and analyse data at the start of any project you undertake; do not leave data collection to the end of the project.

Be clear from the outset what information you need and why so that you can plan your data collection systems accordingly with a clear rationale for your monitoring activities and to keep your efforts focused.

Robust, well-thought out M&E processes will mean that the partnership can better understand what is working, what isn't and ways to address challenges that arise. The information that your M&E system yields will be: a tool for programme and partnership development, data to back up advocacy activities, and to raise the awareness of your work with key stakeholders.

Assumptions

- Data is accessible and of adequate quality to demonstrate progress, understand successes and challenges
- Staff understand the importance of data collection, management, and analysis
- Staff are willing to undertake monitoring and evaluation tasks
- Staff reflect on findings from the data to review practices and implement change where it's needed
- There is resource to transform data into information that can be used to engage with stakeholders
- There is an appetite to engage with stakeholders with findings from institution data
- The institution fosters a culture of learning

Mitigations

- Include exploration and discussion of data accessibility in the planning phase of the project. Where data is missing, establish a means to gather the data or agree proxy measures.
- Gain consensus for data collection tools, especially if introducing a new tool and wherever possible, use existing data collection systems/tools
- Decide on what data is actually needed, and limit collection to that
- Include training on data collection, management and analysis in the project plan. Seek out individuals willing to champion the importance of data
- Plan for regular project meetings that include data review and action components
- Discuss who your stakeholders are, what they want to know about the project, and how best to provide them with this information e.g. in a project meeting, a report, a poster, etc

For more information on evaluation and learning, see Section 7 of the Donation Toolkit.

THET has tools and guidelines for health partnerships to assist them with monitoring and evaluation. See http://www.thet.org/health-partnership-scheme/resources for details.

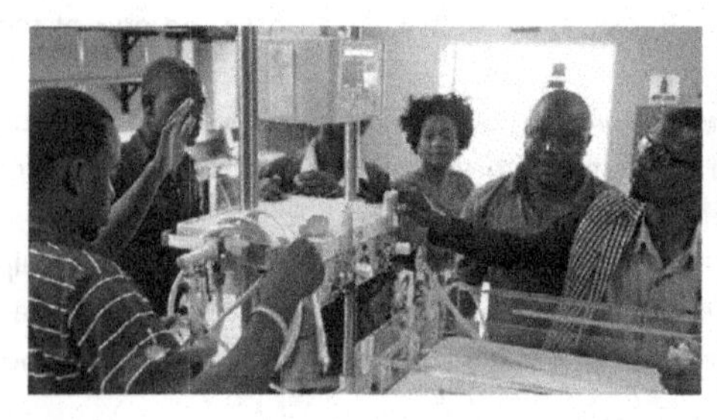

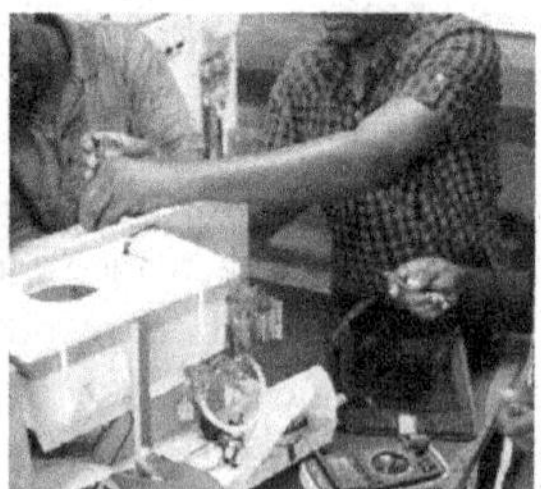

Reporting for BMEs

In general, low- and middle-income countries struggle to procure, manage and maintain medical equipment. This is due to many factors, not least the lack of training and education opportunities for technicians and a lack of spare parts (and consumables). Part of the solution to these two challenges is to collect data. When technicians can prove there is a work overload and a structural lack of spare parts there is a chance that directors and Ministries of Health will become more aware, and will create budgets/priority for solutions; solutions like training people and facilitating access to spare parts.

The way to collect data is well described in the 6 HTM guides we follow in this resource. Some elements are creating, updating and archiving an equipment inventory and equipment history files, which contain manuals, acceptance log sheets, planned preventive maintenance plans and work orders (to know the number of breakdowns and fixes or equally if it is not possible to fix due to lack of spare parts, and to be able to track the equipment through its lifetime. An example of a work order can be found in Ziken's Guide 4 p208.

In general technicians do not like paperwork and prefer to work with tools and equipment. However, the relevance of these types of documents to technicians is that it gives them the opportunity to create a monthly report, which they can present to the hospital director to give visibility to their work, successes and struggles. In Rwanda, working on the administration side of the BMET job has proven very successful and many cases of improvement of status and success have been reported.

Additional resources

Guide 1: How to Organize a System of Healthcare Technology Management

Guide 2: How to Plan and Budget for Healthcare Technology

Guide 3: How to Procure and Commission your Healthcare Technology

Guide 4: How to Operate your Healthcare Technology Effectively and Safely

Guide 5: How to Organize the Maintenance of your Healthcare Technology

Guide 6: How to Manage the Finances of your Healthcare Technology Management Team

http://resources.healthpartners-int.co.uk/resource/how-to-manage-series-for-healthcare-technology/

WHO resources http://www.who.int/medical_devices/management_use/en/

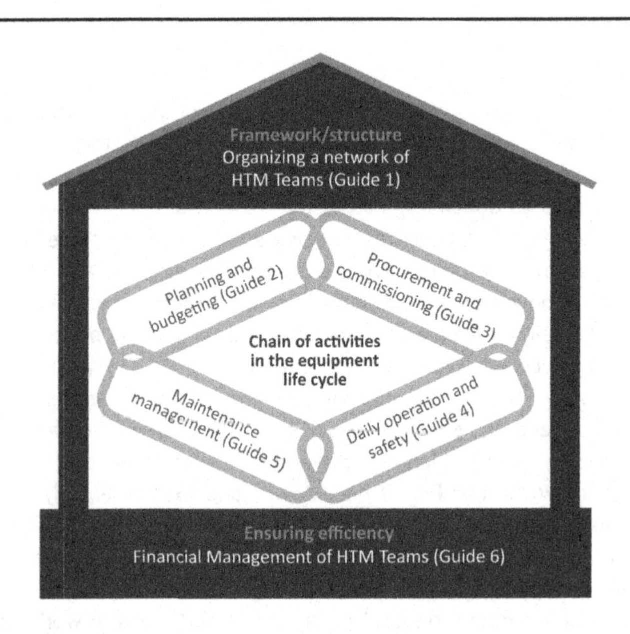

About the Author

Anna Worm is a biomedical engineer focused on training and equipment management in low- resource settings.

With an MSc in BioMedical Engineering from Delft University of Technology (the Netherlands) Anna set up a BSc in BME in Ghana at Valley View University (2007–2008), then joined Philips Healthcare Interventional X-ray headquarters in the Netherlands (2008–2011) before returning to Africa to become Country Manager for Engineering World Health in Rwanda (2011–2013), where she successfully ran a BMET diploma programme. Since the end of 2013 Anna has worked as an independent Biomedical Engineering Consultant for THET. Anna Lives in Benin, West-Africa.

THET is also grateful to the following reviewers; Andrew Gammie, Fishtail Consulting Ltd, Robert Ssetikoleko - part-time lecturer at Makerere University, Kampala, Uganda, Billy Teninty, Marc Myszkowski and Peter Cook - Clinical Engineer at Guy's & St Thomas' Trust in London

This publication was funded through the Health Partnership Scheme, which is funded by the UK.

Department for International Development (DFID) for the benefit of the UK and partner country health sectors and is managed by THET.

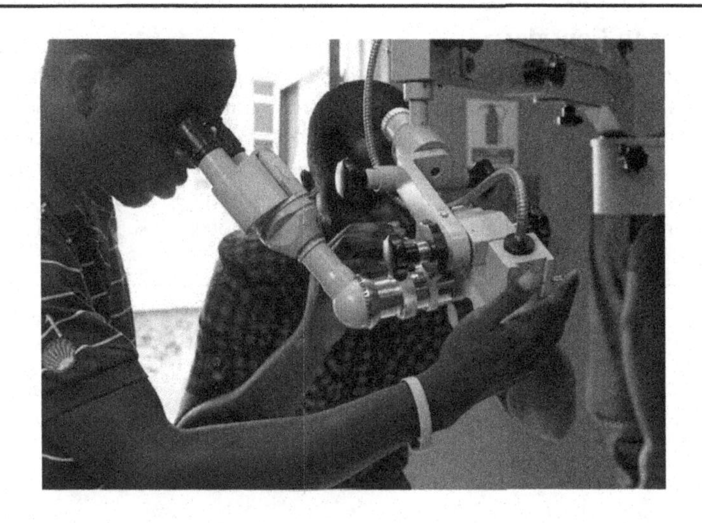

Source: *Used with permission from Worm, Anna, Managing The Lifecycle Of Medical Equipment, Tropical Health and Education Trust (THET), London 2015. Copyright © THET 2015*

Note: All references to page numbers in the "Donations Tool Kit" refer to https://www.thet.org/resources/making-work-toolkit-medical-equipment-donations/

Questions to Consider

The primary focus of this publication is to explore the support of medical devices in low-resource settings. How does the influx of donated equipment to developing nations help or hinder their efforts to deliver safe and effective patient care? This document can offer the reader insights into the challenges of donated equipment support and a lack of trained support staff. Explore the global efforts to visit hospitals, and train staff to manage equipment. Identify groups that both prepare equipment for shipment and conduct courses to guide technician training.

Index

Note: Page numbers followed by "*f*" and "*t*" refer to figures and tables, respectively.

D

E

F

I

Q

R

S

Z

Printed in the United States
By Bookmasters